Fundamentals of Symmetry and Group Theory with Applications in Chemistry

SUJIT SANKAR PANJA
Associate Professor
Department of Chemistry
National Institute of Technology Durgapur, West Bengal

PHI Learning Private Limited
Delhi-110092
2026

*In fond memory of **Shri Asoke K. Ghosh** (October 1942 – February 2024), Founder Chairman and Managing Director of PHI Learning, whose vision endlessly inspires.*

The Legacy Continues....

Published by Pushpita Ghosh, PHI Learning Private Limited, Rimjhim House, 111, Patparganj Industrial Estate, Delhi-110092 and Printed by Syndicate Binders, A-20, Hosiery Complex, Noida, Phase-II Extension, Noida-201305 (N.C.R. Delhi).

₹550.00

FUNDAMENTALS OF SYMMETRY AND GROUP THEORY WITH APPLICATIONS IN CHEMISTRY
Sujit Sankar Panja

ISBN-978-93-5443-879-0 (Print Book)
ISBN-978-93-5443-718-2 (e-Book)

To

My Parents

Mr. Prasanta K. Panja and Mrs. Bharati Panja

Contents

Preface

The study of symmetry and group theory plays a crucial and fundamental role in understanding the mathematical framework behind chemical structures and properties. Symmetry, an inherent characteristic of molecular and crystalline structures, provides essential insights into physical and chemical properties such as bonding, spectroscopy, and reactivity. Group theory, on the other hand, offers a systematic and quantitative approach which allows chemists to predict molecular behavior and simplify complex calculations in chemical bonding, quantum chemistry, molecular vibrations, and electronic transitions.

This book, *Fundamentals of Symmetry and Group Theory with Applications in Chemistry*, is designed to serve as a comprehensive guide for undergraduate and especially postgraduate students, as well as professionals seeking to deepen their understanding of these concepts. It provides a step-by-step introduction to symmetry elements and operations, point groups, representations, and their applications in various chemical contexts. The book emphasizes both theoretical foundations and practical applications, without delving into rigorous or overly complex mathematical formalism—while still highlighting its significance in areas such as spectroscopy, molecular orbital theory, and more.

Each chapter is structured to build upon the previous one, starting with fundamental concepts and gradually advancing to more complex applications. Lots of examples have been presented to illustrate every chapter. Furthermore, the book bridges the gap between abstract mathematical concepts and their tangible chemical applications, making it accessible to a broad audience with varying levels of mathematical background. This book is comprised of 12 chapters, out of which, in Chapters 1–5, the basic concept of symmetry and group theory have been discussed. The Chapters 6–10 discuss several applications in mode of vibrations of molecule, IR and Raman activities, selection rules, electronic transitions and chemical bonding. In Chapter 11, the concept of projection operation technique has been introduced and several applications in chemistry have been discussed. In the last chapter (Chapter 12), the idea of stereographic projection as well as its short application has been presented for students.

I hope that this book will serve as a valuable resource for those exploring the fascinating interplay between symmetry and chemistry, all within a concise volume. Most importantly, I have aimed to present the language and concepts of symmetry and group theory in a simple and accessible manner, so that even those new to the subject may find it engaging. Dedicating my 18 years of teaching experience, I have endeavored to compile my lecture notes into this book, with the goal of making these topics both understandable and inspiring.

I am grateful to the many individuals who have helped me in various ways and contributed significantly to making this book better than it might otherwise have been. I must not forget my colleagues and staff members (Prof. A.K. Patra, Prof. S.C. Moi, Prof. R.N. Saha, Prof. D. Sukul, Prof. M. Maji, Dr. U. Adhikari, Dr. T.K. Saha, Dr. P.N. Chatterjee, Dr. S. Ghosal, Dr. D.R. Banerjee, Dr. S. Laha, Dr. P.K. Ghosh and Swarup Samanta) whose constructive criticisms, suggestions and vibrant environment helped me a lot to gather energy to work and to give a present shape to this book. I would like to thank specially my colleague, Dr. K. Pal, and my sister, Dr. Lopamudra Palodhi, for their unconditional support during the preparation of this manuscript. They were always there whenever I needed help. Through this process, they taught me LaTeX with the patience and care that one would show to a child.

Most importantly, I must acknowledge my students, whose curiosity and encouragement inspired me to write this book in a simple and accessible way. I also extend my deepest respect and gratitude to all my teachers, whose excellent lectures sparked my interest in this subject. In particular, I would like to express special thanks to Prof. A. K. Mukherjee, who first ignited my passion for this topic.

I wish to convey my heartfelt appreciation and love to my wife, Kakoli, and my son, Shinjan, for providing me with an encouraging, peaceful, and joyful environment at home. Their unwavering support and patience throughout the entire journey of preparing the manuscript of this book is unforgettable to me. I am also delighted to mention my sister, Shikha, and my nephew, Anamitro, whose curiosity and interest in the progress of this book brought me added motivation and joy.

Finally, I am sincerely grateful to PHI Learning Private Limited for considering and agreeing to publish my manuscript in the form of a book.

In spite of our sincere effort to keep the book error free, still if the readers find any, and they point out such mistakes and share constructive suggestions for improvement of this book, it would be highly appreciated.

Sujit Sankar Panja

Chapter 1

Symmetry Elements and Symmetry Operations

1.1 Knowledge of Symmetry is Important

To demonstrate how some physical properties of a body get organized or reorganized with the change of symmetry, consider the potential energy levels of three blocks of different dimensions.

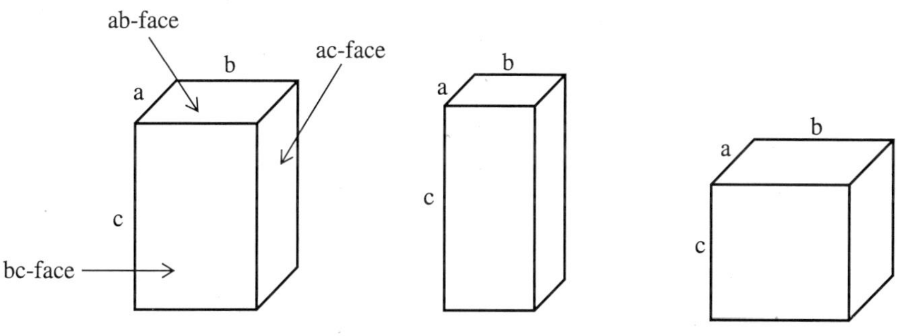

Figure 1.1: Three solid blocks of different dimensions on different resting faces.

Each of these solid blocks has six stable resting faces in the earth's gravitational field *viz*., two *ab*, two *ac* and two *bc* faces. The potential energy (PE) V, associated with each resting face, can be calculated from the well-known formula, $V = mgh$, where m is the mass of the block, g is acceleration due to gravity and h is the height of the centre of mass (COM) above the surface on which the object rests.

The potential energies of the six resting positions of the block-*I* are divided into three groups, depending on which face it rests. The potential energy is calculated to be highest when its centre of mass is at highest position and conversely, it is lowest when COM is lowest *i.e.,* when it rests on *ab* face, PE is maximum and at it's *bc* face,

PE is minimum. The block-*II* has two energetically distinct resting faces because of the equal length of side *a* and *b*. All six faces of block-*III* are equally probable and orientations are of equal energy, *i.e.* degenerate. Hence, in this example, it is evident that as the symmetry of the solid blocks increases in the order $I < II < III$, the number of dissimilar energy levels decreases in the same order (Figure 1.2). The degeneracy of each level increases with increasing symmetry. Here, it should be noted that the energy levels are not in according to scale in the picture.

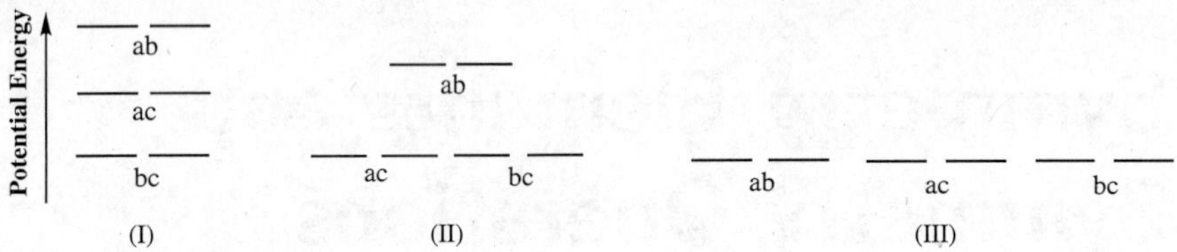

Figure 1.2: Potential energy levels of three blocks of different shapes (not in scale).

The same trend is true for molecules as well. The more symmetric the molecule is, the fewer different energy levels it possesses and the greater degeneracy is observed. The study of symmetry, therefore, helps us to simplify some problems by reducing the number of different probable parameters. Therefore, in that case we do not need to calculate all the different possible unknown parameters, since knowledge of symmetry depicts that they will have the similar values for those parameters. It is also possible to predict the symmetry of a particular vibrational mode of a molecule from a knowledge of symmetry and group theory. Symmetry also helps us to decide which spectroscopic transitions are feasible and which transitions are forbidden among different energy states. From symmetry point of view, it is also possible to predict proper hybridization of central atom as well as the structure of a molecule. There are several others these kind of aspects which can be addressed from symmetry point of view.

Let us take a real example of molecules and see how energy levels reorganize with change in symmetry. In the following table, CH_4, CH_3D and CH_2D_2 molecules are in order of decreasing symmetry. At this moment, do not bother about the notation used in parentheses with the molecules in the table. We can see from Table1.1, that degeneracy increases or degrees of freedom decreases going from less symmetry to more symmetry, which is evident from the number of vibrational frequencies for each molecule.

Although, it is too early to discuss this particular interesting example of very basic application of symmetry in determining the direction of dipole moment of a molecule, but still I could not resist myself in doing so. At this moment, do not bother about the notations/symbols (symmetry operation) used here. But once this chapter is completed, you are suggested to come to this point again and I am sure you will love to read this part.

The basic idea of molecular symmetry is based on a simple fact that *no symmetry operation can change an intrinsic molecular physical property*. For example, when a molecule is flipped or rotated 180 degrees through one suitable axis (which is called a C_2 operation), it has the following two equivalent meanings:

1. *The observer rotates the molecule clockwise through 180 degrees about an suitable axis and gets an equivalent configuration.* Or,
2. *The molecule is kept fixed and the observer himself turns around the same axis anticlockwise through 180 degree and then observe the molecule.*

Table 1.1: Different vibrational frequencies (cm^{-1}) of CH_4, CH_3D and CH_2D_2

CH_4	$\nu_1(A_1)$	$\nu_1(E)$		$\nu_3(T_2)$		$\nu_4(T_2)$		
(T_d)	2914	1526		3020		1306		
CH_3D	$\nu_2(A_1)$	$\nu_5(E)$		$\nu_1(A_1)$	$\nu_4(A_1)$	$\nu_3(A_1)$	$\nu_6(E)$	
(C_{3v})	2205	1477		2982	3030	1306	1156	
CH_2D_2	$\nu_2(A_1)$	$\nu_3(A_1)$	$\nu_5(A_2)$	$\nu_1(A_1)$	$\nu_6(B_1)$	$\nu_8(B_2)$	$\nu_4(A_1)$	$\nu_7(B_1)$	$\nu_9(B_2)$
(C_{2v})	2139	1450	1286	2974	3030	2255	1034	1090	1235

* Data from K. Nakamoto, Infrared spectra of inorganic and coordination compounds, 2nd ed., John Wiley & Sons, New York, 1970, p.113.

Obviously, a physical property, such as the dipole moment, must not change as a result of 180 degree rotation through one suitable axis (Here, C_2-axis). Based on this logic, we can predict the existence or non-existence of dipole moment in a molecule. For that, first it is assumed that the given molecule has some dipole moment vector in some arbitrary chosen direction. Then all the possible symmetry operations present for the molecule are applied and examined whether the assumed dipole moment vector changes or not. If the vector remains unchanged under the action of all the symmetry operations of the molecule, then that dipole moment vector really exist, otherwise it does not. Let us take few examples to illustrate this idea.

1.2 Determination of Dipole Moment of a Molecule from the Knowledge of Symmetry

1.2.1 Benzene Molecule (D_{6h}):

Let us imagine, that benzene molecule has a dipole moment (μ) in any arbitrary direction, say along C_6 axis, as shown in Figure 1.3. The list of symmetry elements for this point group are as follows:

$$E, C_6, \sigma_h, C_2(\text{perpendicular to } C_6), \sigma_v, \sigma_d, i, S_3 \text{ and } S_6$$

Now, let us check the effect of *symmetry operations* (after a few moments, we will discuss this term) on μ:

$$E(\mu) = \mu;\ \ C_6(\mu) = \mu;\ \ \sigma_v(\mu) = \mu$$

Here we see that under some operations μ remains unchanged whereas, under some other operations, μ changes its direction. *e.g.*,

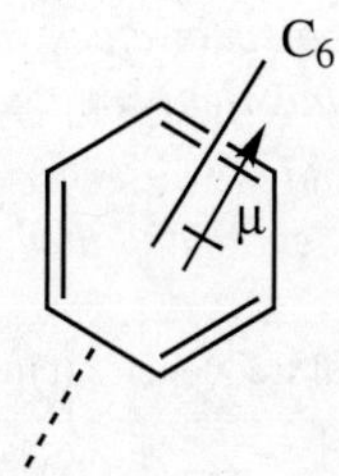

Figure 1.3: The principal C_6 axis and arbitrary dipole moment vector (μ) of Benzene molecule.

$$i(\mu) = -\mu;\ C_2(\mu) = -\mu;\ \sigma_h(\mu) = -\mu$$

Since, under some operations, μ changes its direction, therefore, considering the basic concept of molecular symmetry it can be concluded that the arbitrary direction of μ, which was assumed, is not correct.

Similar way, μ can be imagined in any other direction on the benzene plane and would be found that the direction of μ changes under few operations. Since, in any direction the presence of dipole moment could not be established, hence this can be concluded that benzene molecule does not possess any dipole moment.

1.2.2 Water Molecule (C_{2v}):

For water molecule, let us imagine, the resultant dipole moment is along one O–H bond, as shown in the figure.It is seen from Figure 1.4 that rotation along C_2 axis and one vertical plane (σ_{xz}) will change the direction of imagined dipole moment in the first case (left hand side). Hence, it can be concluded that this imaginary direction is not correctly guessed.

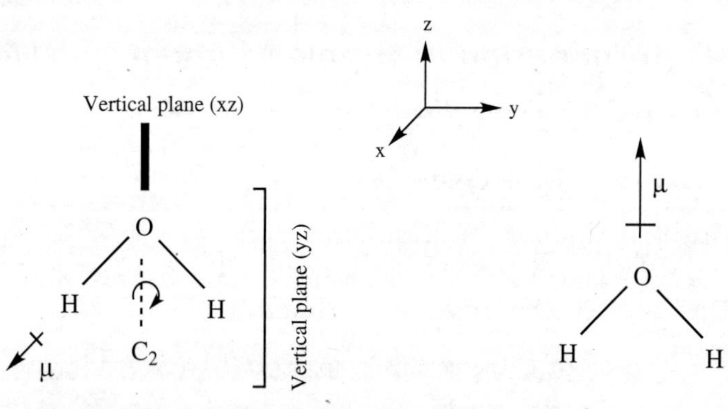

Figure 1.4: Two different imaginary directions of dipole moment of water molecule.

Now, let us take the second imaginary direction (right hand side) of dipole moment of water molecule. Here we can see that all the four symmetry operations $(E, C_2, \sigma_{xz}, \sigma_{yz})$ present for water molecule will retain the direction of the proposed dipole moment. Hence, this proposed direction of dipole moment is the correct one. Also we can take the example of $CHCl_3$ and CCl_4 and try similar type of exercise.

From these examples discussed above, we see that there are various interesting applications of molecular symmetry in chemistry.

Now, let us discuss the meaning of symmetry elements and symmetry operations.

1.3 Symmetry Elements and Symmetry Operations

Generally, we speak of a shape as "symmetrical" and "unsymmetrical" or "more symmetrical. For scientific purpose however, we need to specify ideas of symmetry in a more systematic and quantitative way. Let us start the discussion by asking a very simple question. Identify the most and least symmetric among A, B or C as listed in Figure 1.5?

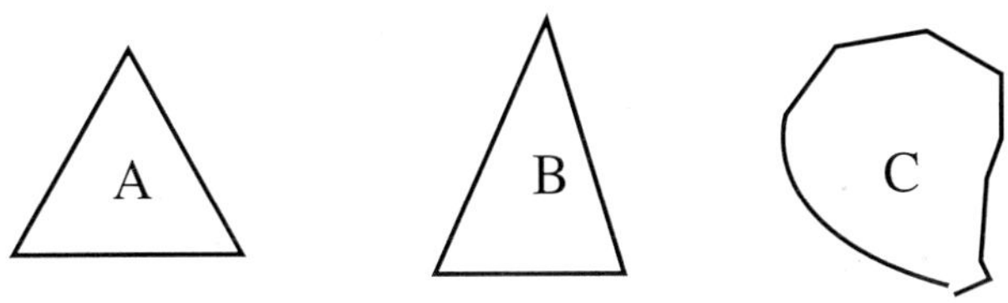

Figure 1.5: Objects of various symmetry.

Of course, A is the most symmetric and the object, C is the least symmetric!! To answer this question, we need not to be an expert in the field of symmetry, this can be answered correctly from our very in-born intuition.

1.3.1 Proper Rotational Axis of Element (C_n)

Let us put the idea of symmetry in a more quantitative way as described in Figure 1.6, where one axis is imagined through the letter A, perpendicular to the plane of the triangle. Now, we will perform two exercises; Firstly, if we rotate the triangle through this axis clockwise 60 degree, we get structure B, which is visually completely different from original one and can easily be distinguished even if we remove the numbering system on the three vertices. Hence, this kind of rotations through any arbitrary angle, may produce a structure which becomes easily distinguishable from the original one by just giving a look, are not considered as symmetry operation. Now have a look on the second exercise (shown in lower panel) where the triangle is rotated clockwise through the same axis by an angle of 120 degree.

Here, in the second case, A and A' are *indistinguishable* but obviously not identical. We say this category of rotation as a *symmetry operation*. Hence, *when after some movement to an object or structure, it produces a similar structure which is indistinguishable from the initial one, is called a symmetry operation*. Indistinguishable may or may not be the original initial structure or identical structure. First, we have to understand the term *indistinguishable* and *identical* clearly. Here identical means the same structure or original structure from where we started the exercise, whereas, indistinguishable means that it is not necessarily be the identical one, but very similar to identical one. In this context, another point must be kept in mind

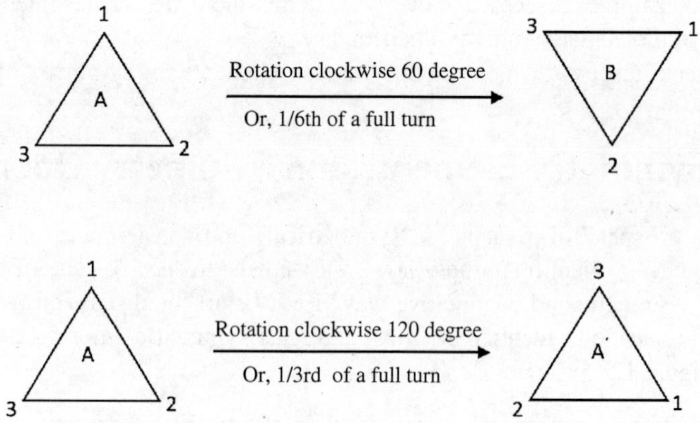

Figure 1.6: Illustration of rotation of a triangle.

that *after symmetry operation, no physical properties (e.g., energy, dipole moment etc.) changes for any particular object or molecule*. We can also think of other types of rotation for the triangle, 'A'. Here the triangle is rotated around the dashed line (passing through vertex 1) through 180 degrees, resulting in interchange of position of vertex 2 and 3, although vertex 1 remains unaltered (Figure 1.7). Since, this type of movement produces indistinguishable structure, hence it is designated as one rotational symmetry operation for this object. This way we can think of other two symmetry rotations along the other two remaining vertices, 2 and 3.

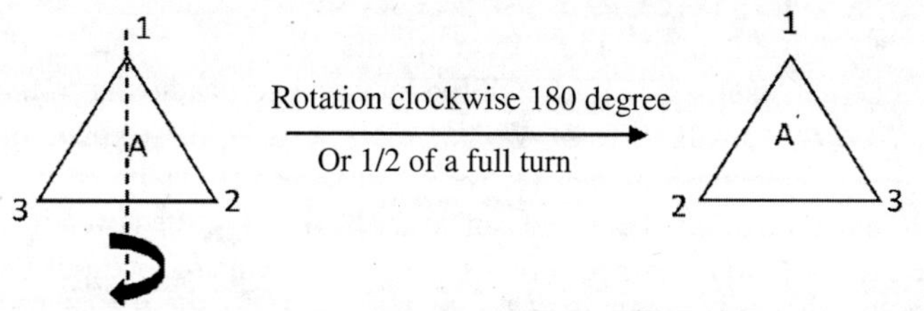

Figure 1.7: Illustration of rotation of a triangle through other possible way.

These kinds of rotations around those axes are called ***Proper rotation*** and are designated by C with a subscript n (*i.e.*, C_n), where n is the number of operations through which we can achieve the identical or initial structure, is called the *order* of rotation. If more than one different type of axes are present for a molecule, the highest order axis is called the ***principal proper rotational axis*** or simply ***principal axis*** of symmetry.

Hence, when we rotate the triangle by 120 degree, the symmetry operation is C_3 and in the second case, when we rotate by 180 degree we designate the operation by C_2.

For a square plane, there is a C_4 proper rotational axis perpendicular to the plane of the object and also four C_2 axes are there through the opposite vertices as well as midpoint of opposite bonds as shown in Figure 1.8. Here C_4 axis is the principal axis of symmetry.

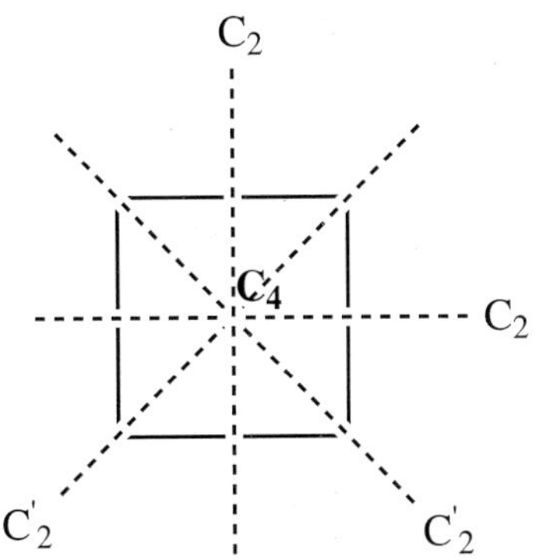

Figure 1.8: Illustration of various rotational axes of a square planar object.

It may be noted that one symmetry element can generate more than one symmetry operations by repetition of similar exercises, shown in Figure 1.9.

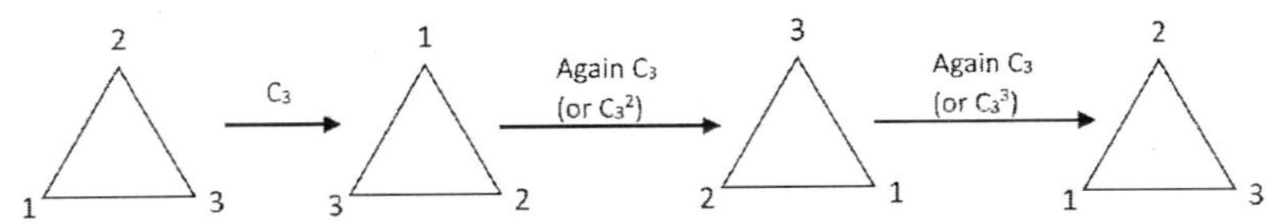

Figure 1.9: Generation of symmetry operations from symmetry element.

1.3.2 Identity Element (E)

We have just seen that C_3 element can generate three symmetry operations, namely C_3, C_3^2 and C_3^3. When a rotation of 120 degree is done, through the centre of triangle, it is called C_3^1 operation or simply the C_3 operation. Another 120 degree rotation in the same direction results a C_3^2 operation from the initial structure and consequently a third time rotation by another 120 degree gives C_3^3 symmetry operation. The interesting point is that the C_3^3 operation generates a structure which is not only indistinguishable, but also identical to the initial structure. Hence, C_3^3 is alternatively denoted by E, ***identity*** operation. Hence, identity operation indicates *doing nothing*, but later we will see that identity symmetry element has very important implications in group theory.

1.3.3 Plane of Symmetry Element (σ)

Role of this plane of symmetry is very similar to reflection on a mirror. Many molecules have one or more than one plane of symmetry elements (σ). This is possessed by all planar molecules. A flat molecule always retains a molecular plane, *e.g.*, in H_2O, Benzene *etc*. Here, since the planes (*xz* and *yz*) are along the principal

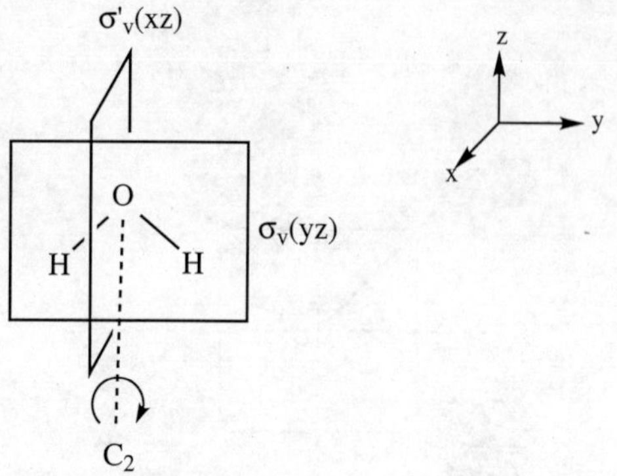

Figure 1.10: Different plane of symmetries for water molecule.

axis hence, both are designated as ***vertical planes*** (σ_v) as shown in Figure 1.10 (The subscript, *v* stands for *vertical*). Although these two are vertical planes, still they are not equivalent which is evident from the results after reflection through these two planes. In one case, the two hydrogen atoms mutually reflect each other and in the other plane, the whole molecule gets reflected. To distinguish themselves from each other, one plane has been designated by indicating a prime($'$).

There is another kind of plane which is termed as ***horizontal plane*** (σ_h) and is perpendicular to the principal axis as shown below. Here C_6 is the principal axis

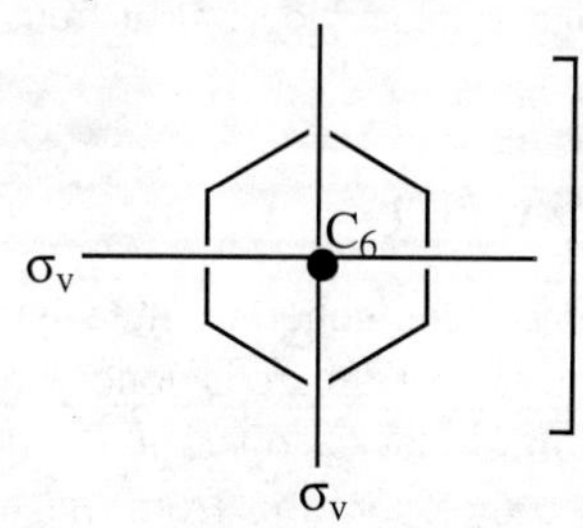

which passes through the plane of the benzene ring perpendicularly. The plane which is coplanar with the benzene molecule and perpendicular to this axis as well, is taken as horizontal plane (σ_h). The other two planes shown, are designated as vertical planes (σ_v), since these two planes contain the principal axis. Therefore, to distinguish between these two types of planes (vertical and horizontal), the prime job is to find out the principal axis of symmetry for a molecule. At this point, it may be noted that, similar to proper rotation,in case of two consecutive reflection operations, also ended up with an *identity operation* (*i.e.*, $\sigma^2 = E$).

Later, taking some example, we will also learn another type of plane, called ***dihedral plane of symmetry*** (σ_d), which basically bisect two C_2 axes.

1.3.4 Inverse Element (i)

Consider a point, P on the C–C bond in the staggered ethane molecule, as shown in Figure 1.11. The point P is such a point from which if we start a journey to equal and in opposite direction, we will reach to an equivalent point. e.g., consider the pair H_1 and H_4 or H_2 and H_5 or H_3 and H_6. This point, P is called the *inversion center* (i) for the molecule. This center may be present on a bond or on an atom or in space (*e.g.*, benzene) as well.

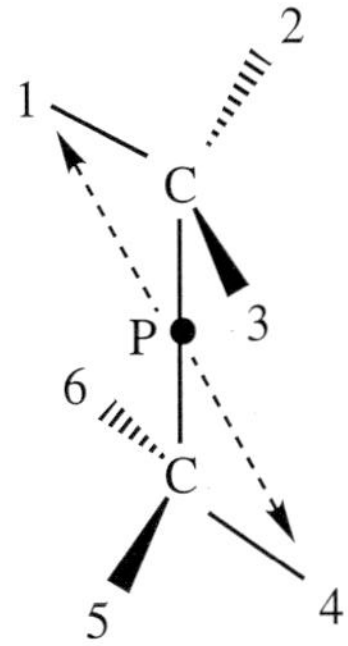

Figure 1.11: Illustration of inversion point (i).

1.3.5 Improper Rotational Axis of Element (S_n)

Improper rotation is an another important symmetry operation, designated by S_n. It is basically not a pure symmetry operation, rather a combination operation consisting of pure rotation by $1/n$ of a turn (C_n axis), followed by reflection in a plane perpendicular to that axis. Molecules possessing an improper axis, can be of two types:

1. *Both, the proper axis (C_n) and the reflection plane, perpendicular to that axis (σ_h) are present.*
2. *Neither the proper axis (C_n) nor reflection plane, σ_h are present, but the improper axis of symmetry (S_n) is present.*

BCl_3 is the example of first kind, where C_3 axis is present along with a plane perpendicular to that axis (σ_h). Here, a rotation through 120 degree followed by reflection on the molecular plane results in an indistinguishable structure. Hence, the molecule is said to own an S_3 axis.

Example of second category is staggered ethane where it has S_6 axis along the C–C bond although, it has **neither** C_6 **nor** σ element of symmetry. The three consecutive S_6 operations have been shown in Figure 1.12.

Similarly, the operations S_6^4, S_6^5, S_6^6 can be completed. It would be very interesting to note that S_6^2 operation is nothing but a C_3 operation. Similarly, S_6^3 operation basically depicts an inversion operation, i , S_6^4 is nothing but C_3^2 operation and S_6^6 is equivalent to identity operation, E. However, S_6^5 operation is a unique operation that can not be represented by any pure alternative operation. Thus, for staggered ethane, the improper rotational axis, S_6 produces S_6 and S_6^5 symmetry operations.

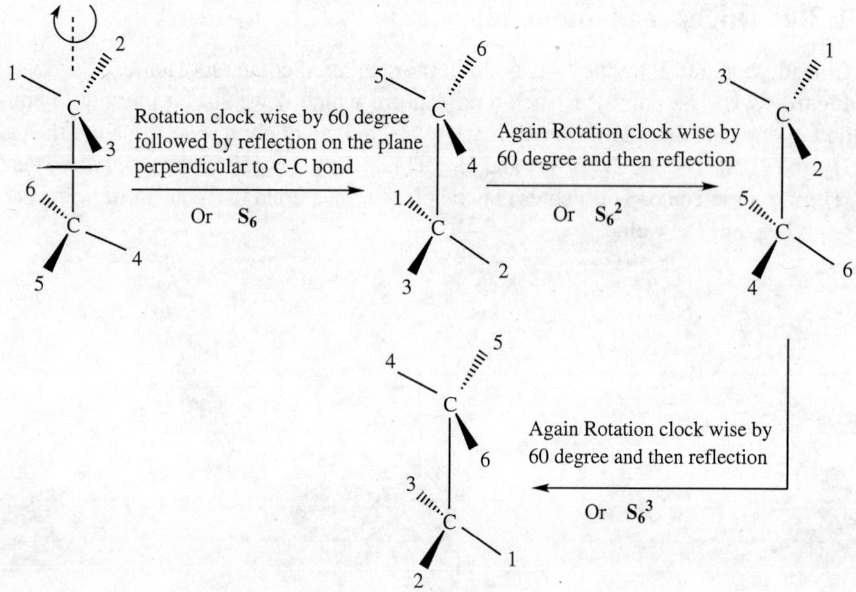

Figure 1.12: Illustration of improper rotational axis.

The other associated operations of S_n axis can also be tried. The number of associated operations of S_n axis depend on the nature of the order, n, whether it is even or odd. Let us try with few examples and see what kind of results we get.

S$_2$:

$S_2^1 = \sigma C_2^1$

$S_2^2 = (\sigma C_2^1)(\sigma C_2^1) = \sigma^2 C_2^2 = EE = E$

S$_3$:

$S_3^1 = \sigma C_3^1$

$S_3^2 = (\sigma C_3^1)(\sigma C_3^1) = (\sigma^2 C_3^2) = E.C_3^2 = C_3^2$

$S_3^3 = (\sigma C_3^1)(\sigma C_3^1)(\sigma C_3^1) = (\sigma C_3^3) = \sigma E = \sigma$

$S_3^4 = (\sigma C_3^1)(\sigma C_3^1)(\sigma C_3^1)(\sigma C_3^1) = EC_3^4 = EC_3^1 = C_3$

$S_3^5 = (\sigma C_3^1)(\sigma C_3^1)(\sigma C_3^1)(\sigma C_3^1)(\sigma C_3^1) = \sigma^5 C_3^5 = \sigma C_3^2$

$S_3^6 = (\sigma C_3^1)(\sigma C_3^1)(\sigma C_3^1)(\sigma C_3^1)(\sigma C_3^1)(\sigma C_3^1) = EC_3^6 = (C_3^3)^2 = E$

S$_4$:

$S_4^1 = \sigma C_4^1$

$S_4^2 = (\sigma C_4^1)(\sigma C_4^1) = (\sigma^2 C_4^2) = E.C_2 = C_2$

$S_4^3 = (\sigma C_4^1)(\sigma C_4^1)(\sigma C_4^1) = \sigma C_4^3$

$S_4^4 = (\sigma C_4^1)(\sigma C_4^1)(\sigma C_4^1)(\sigma C_4^1) = EC_4^4 = E.E = E$

S_5:

$S_5^1 = (\sigma C_5^1)$

$S_5^2 = (\sigma C_5^1)(\sigma C_5^1) = EC_5^2 = C_5^2$

$S_5^3 = (\sigma C_5^1)(\sigma C_5^1)(\sigma C_5^1) = \sigma C_5^3$

$S_5^4 = (\sigma C_5^1)(\sigma C_5^1)(\sigma C_5^1)(\sigma C_5^1) = EC_5^4 = C_5^4$

$S_5^5 = (\sigma C_5^1)(\sigma C_5^1)(\sigma C_5^1)(\sigma C_5^1)(\sigma C_5^1) = (\sigma C_5^5) = \sigma E = \sigma$

$S_5^6 = (\sigma C_5^1)(\sigma C_5^1)(\sigma C_5^1)(\sigma C_5^1)(\sigma C_5^1)(\sigma C_5^1) = EC_5^6 = C_5^1$

$S_5^7 = (\sigma C_5^1)(\sigma C_5^1)(\sigma C_5^1)(\sigma C_5^1)(\sigma C_5^1)(\sigma C_5^1)(\sigma C_5^1) = (\sigma C_5^7) = \sigma C_5^2$

$S_5^8 = (\sigma C_5^1)(\sigma C_5^1)(\sigma C_5^1)(\sigma C_5^1)(\sigma C_5^1)(\sigma C_5^1)(\sigma C_5^1)(\sigma C_5^1) = EC_5^8 = C_5^3$

$S_5^9 = (\sigma C_5^1)(\sigma C_5^1)(\sigma C_5^1)(\sigma C_5^1)(\sigma C_5^1)(\sigma C_5^1)(\sigma C_5^1)(\sigma C_5^1)(\sigma C_5^1) = (\sigma C_5^9) = \sigma C_5^4$

$S_5^{10} = (\sigma C_5^1)(\sigma C_5^1)(\sigma C_5^1)(\sigma C_5^1)(\sigma C_5^1)(\sigma C_5^1)(\sigma C_5^1)(\sigma C_5^1)(\sigma C_5^1)(\sigma C_5^1)$

$= EC_5^{10} = (C_5^5)^2 = E$

S_6:

$S_6^1 = \sigma C_6^1$

$S_6^2 = (\sigma C_6^1)(\sigma C_6^1) = EC_6^2 = C_3$

$S_6^3 = (\sigma C_6^1)(\sigma C_6^1) = \sigma C_6^3 = i$

$S_6^4 = (\sigma C_6^1)(\sigma C_6^1)(\sigma C_6^1)(\sigma C_6^1) = EC_6^4 = C_3^2$

$S_6^5 = (\sigma C_6^1)(\sigma C_6^1)(\sigma C_6^1)(\sigma C_6^1)(\sigma C_6^1) = \sigma C_6^5$

$S_6^6 = (\sigma C_6^1)(\sigma C_6^1)(\sigma C_6^1)(\sigma C_6^1)(\sigma C_6^1)(\sigma C_6^1) = EC_6^6 = E$

Thus, from the results it can be concluded that S_n axis is associated with $2n$ number of symmetry operations, when n is odd and is associated with n number of symmetry operations, when n is even. It must be remembered that all the associated operations do not produce unique operation, rather most of the associated operations can be represented by some other alternative simple symmetry operation. For example, S_3^5 operation ended up with the binary operation, (σC_3^2); it is a unique operation, but S_3^4 or S_3^6 can be replaced by C_3 or E respectively. Hence, we do not need to count the contribution of S_3^4 or S_3^6.

Let us see what are the different symmetry operations present in NH_3 molecule (Figure 1.13). The list of symmetry operations are: E, C_3, C_2^3, σ_v, $\sigma_{v'}$ and $\sigma_{v''}$.

1.4 Combination of Two Symmetry Operations

Already we have gathered some experience of binary operation during the discussion of improper rotation. To illustrate combination of two symmetry operations (Fig. 1.14), let us take H_2O molecule as our model with one H-atom labelled with an arrow. Upward arrow indicates, above the plane of molecule or paper and downward arrow indicates below the plane of the molecule. Let us perform the binary operation, $(\sigma_{xz}\ C_2)$; where, C_2 is taken as first operation and σ_{xz} as second operation. It

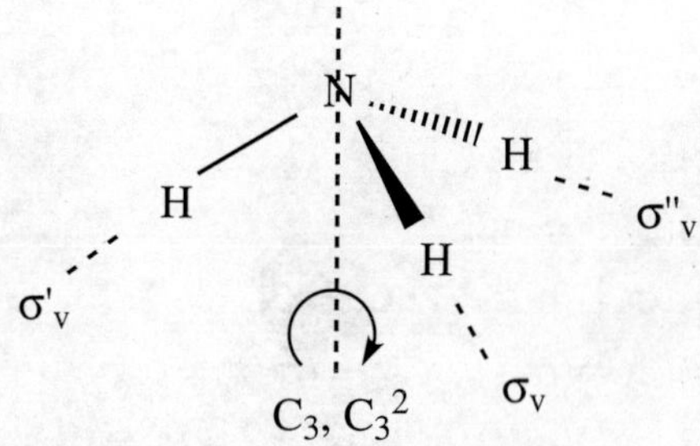

Figure 1.13: Various symmetry elements present in NH_3 molecule.

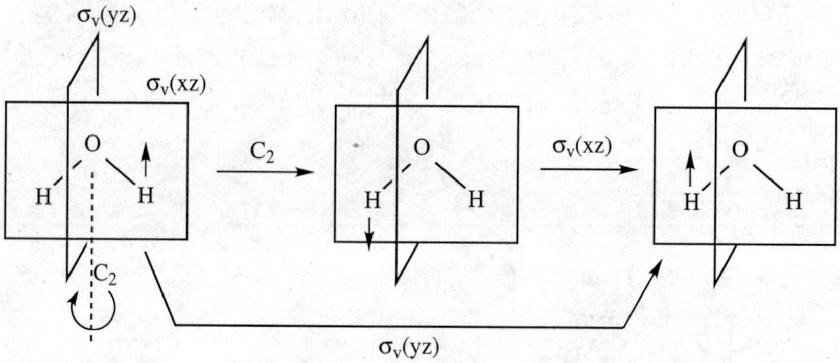

Figure 1.14: Illustration of binary symmetry operation.

is found from the figure that the result of these two consecutive operations can be alternatively represented by simple σ_{yz} or σ' operation.

Therefore, it can be written as; $(\sigma_{xz}\ \mathrm{C}_2) = \sigma_{yz}$

Now, let us see what happens if we reverse the order of the operation, *i.e.*, $\mathbf{C}_2\ \sigma_{xz}$. Again, in a similar manner we get $\mathbf{C}_2\ \sigma_{xz} = \sigma_{yz}$. Hence, it can be concluded that $\sigma_{xz}\ \mathbf{C}_2 = \mathbf{C}_2\ \sigma_{xz}$ ***i.e.*, these two operations commute**. Similarly, we can exercise the other binary combination operations by taking every possible pair of elements present in $\mathrm{C}_{2}v$ point group and a *multiplication table* can be constructed. It should be mentioned that the first row and first column of the table designate the symmetry operations as well as the name of a *point group*. The final form of the multiplication table would be as follows, shown in Table 1.2.

Here, the notable observation is that every element commutes with every other elements of the group. When this kind of situation arises, the set of elements or group is called *abelian* group . But please note that this is not going to be true for every point group.

There are other groups of relatively larger size where some other additional features can be observed. C_{3v} point group (**NH_3**) is such an example of this kind, containing symmetry operations, E, C_3, C_3^2, σ_v, σ_v' and σ_v''. In a similar way, we can try to construct thc 6×6 multiplication tablc for this group. To construct this, we can take a projection of ammonia molecule on a plane where one point is considered (P) and observed the effect of various operations. The C_3 element passes through

Table 1.2: Multiplication table of water molecule (C_{2v})

1st operation →

2nd operation ↓

C_{2v}	E	C_2	σ_{xz}	σ_{yz}
E	E	C_2	σ_{xz}	σ_{yz}
C_2	C_2	E	σ_{yz}	σ_{xz}
σ_{xz}	σ_{xz}	σ_{yz}	E	C_2
σ_{yz}	σ_{yz}	σ_{xz}	C_2	E

N-atom perpendicular to the plane of the paper and various planes are shown in the projected diagram. Let us start with a binary operation $\sigma'_v\ C_3$, where C_3 is the first operation and σ'_v is the second operation. It is found that the result of the binary operation can be easily replaced by a single alternative operation namely σ''_v. So we can write $\sigma'_v\ C_3 = \sigma''_v$.

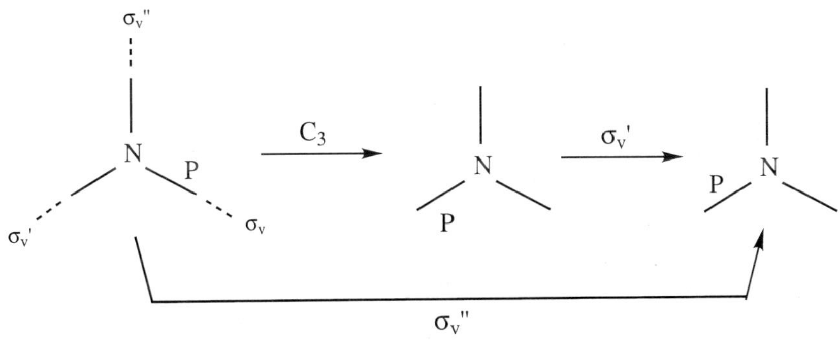

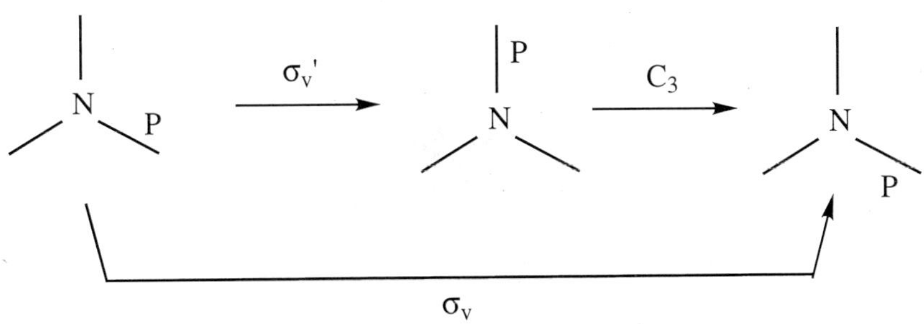

But now, if we perform the reverse order of binary operation, *i.e.*, $C_3\sigma'_v$, we do not find σ''_v, rather we get σ_v, which is evident from the diagram below. Hence, $C_3\sigma'_v = \sigma_v$. At this point, the similar case for C_{2v} point group may be recalled. So here interesting notable point is that $\boldsymbol{\sigma'_v C_3 \neq C_3\sigma'_v}$, *i.e.*, σ'_v and C_3 do not commute. If we try all other possible binary combinations, it is found that E commutes with all other elements and C_3 commutes with C_3^2. But interestingly, three σ_v's do not commute with each other.

In a similar way we find out other possible binary combinations and can complete the multiplication table.

Table 1.3: Multiplication table of NH_3 molecule (C_{3v} point group)

$\mathbf{C_{3v}}$	**E**	$\mathbf{C_3}$	$\mathbf{C_3^2}$	σ_v	σ_v'	σ_v''
E	E	C_3	C_3^2	σ_v	σ_v'	σ_v''
C_3	C_3	C_3^2	E	σ_v'	σ_v''	σ_v
C_3^2	C_3^2	E	C_3	σ_v''	σ_v	σ_v'
σ_v	σ_v	σ_v''	σ_v'	E	C_3	C_3^2
σ_v'	σ_v'	σ_v	σ_v''	C_3	C_3^2	E
σ_v''	σ_v''	σ_v'	σ_v	C_3^2	E	C_3

1.5 Few Examples

Let us see few examples of different symmetry operations present in few common molecules:

1.5.1 H_2O Molecule

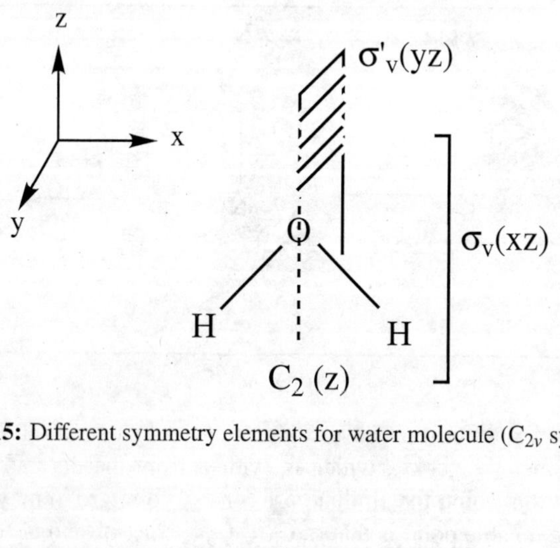

Figure 1.15: Different symmetry elements for water molecule (C_{2v} symmetry).

1.5.2 BCl_3 Molecule:

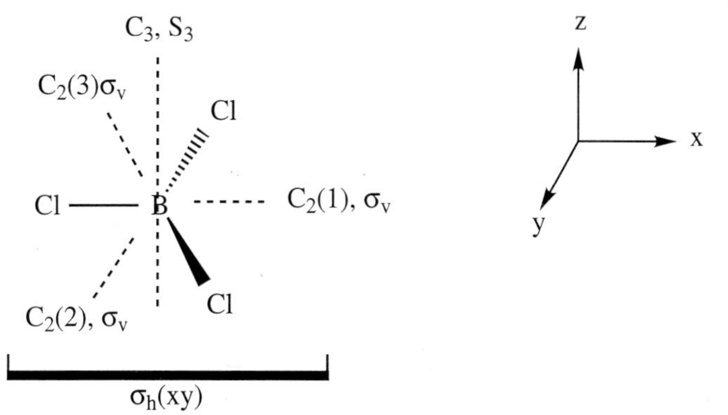

Figure 1.16: Different symmetry elements for BCl_3 molecule (D_{3h} symmetry).

1.5.3 AB_4 Type of Molecule (Tetrahedral Geometry):

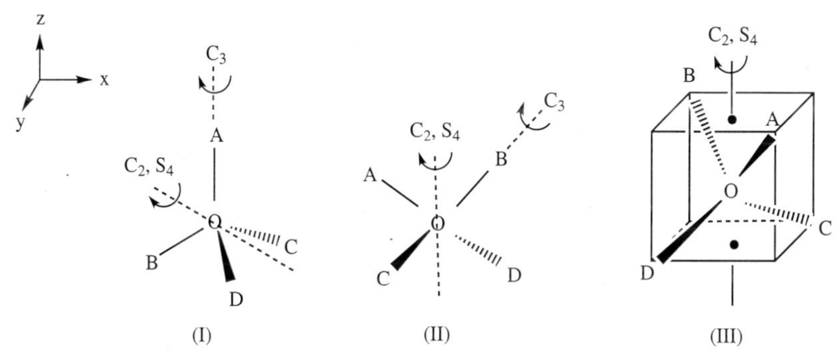

Figure 1.17: Different orientations of a tetrahedral, AB_4 molecule along with different symmetry elements.

In T_d, the symmetry operations are $E, 4C_3, 4C_3^2, 3C_2, 3S_4, 3S_4^3$, and $6\sigma_d$. Some important features of a tetrahedron:

- O is the centre of the T_d in space.
- There are four vertices (A, B, C and D); four C_3 axes passing through four vertices and central atom as well.
- There are 4 trigonal planes (ABD, ABC, BCD, ACD).
- There are 6 edges (line between two vertices or corners where atoms are present; AB, AC, AD, BD, BC and CD).
- C_2-axis passes through the mid-point of each edge or mid-point of each face. Hence, there are three such C_2's namely $C_2(x)$, $C_2(y)$ and $C_2(z)$.

- S_4 axis is collinear with C_2. For each S_4 axis, there will be two unique operations, S_4^1 and S_4^3. It may be noted that S_4^2 and S_4^4 are not considered as they are equivalent to C_2 and E operations respectively. There are total such three S_4 axes present collinear with three C_2 axes .
- σ_d is the plane bisecting two C_2 axes and passing through each edge. Since, there are such six edges, hence, there would be total six such planes.

Few symmetry operations are shown in Figure 1.18.

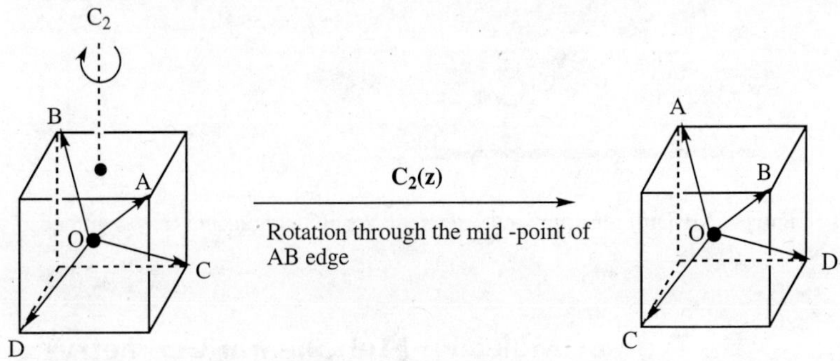

[Similarly, other two axes are through AC edge and AD edge corresponds to $C_2(x)$ and $C_2(y)$ axis, respectively. There are total such THREE C_2-axis].

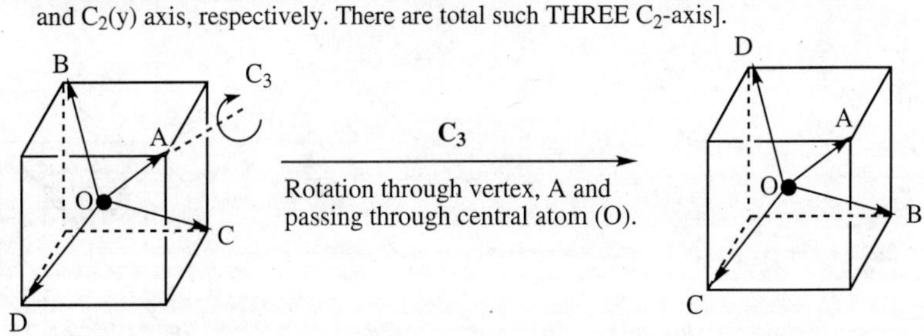

[Similarly, through other three vertices (B, C and D) and central atom,O. There are total FOUR such C_3 axes].

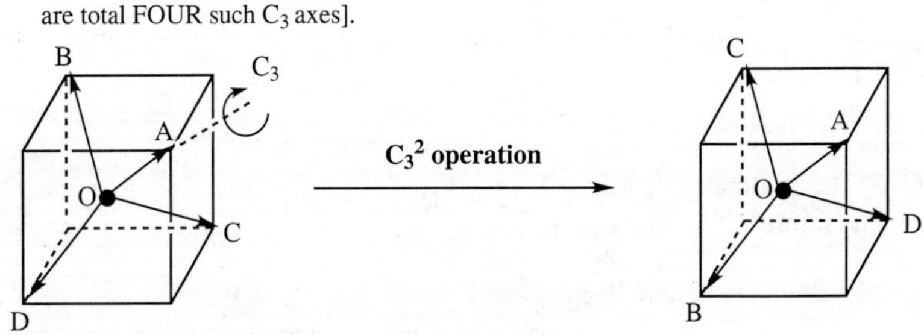

[There are total EIGHT operations through four C_3 axes ($4C_3^1$ and $4C_3^2$)].

Figure 1.18: (*Contd.*)

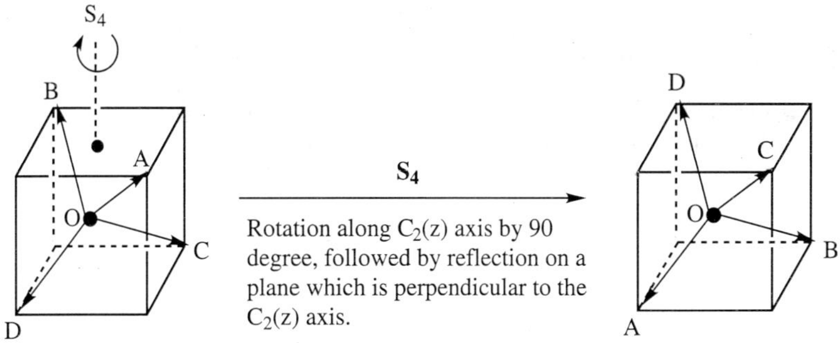

[Colinear with the three C_2 axes. (Here along $C_2(z)$ is shown). There are total THREE S_4 axes along three C_2 axes. Each S_4 axis produces two operations, S_4^1 and S_4^3, giving total SIX operations].

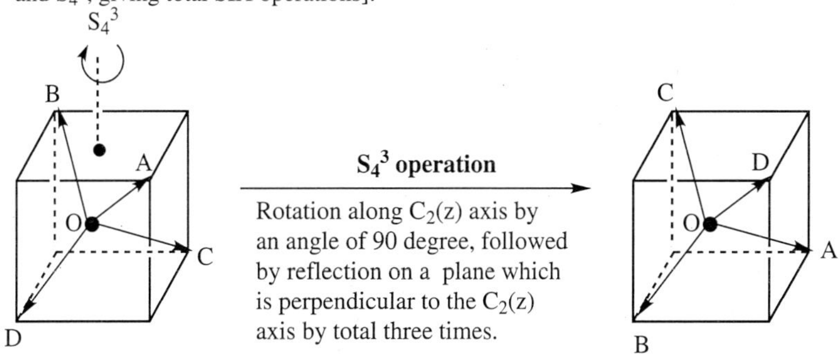

[Along S_4 axis. (Here along $C_2(z)$ is shown)].

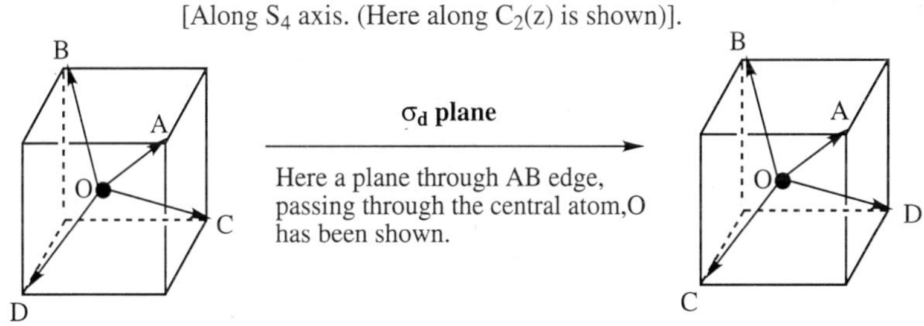

[This plane is bisecting two C_2's and passing through each edge. There are one edge on each face connecting two corresponding vertices where atoms are present. Other five edges are through AC, AD, BC, BD and CD faces. There are total SIX dihedral planes].

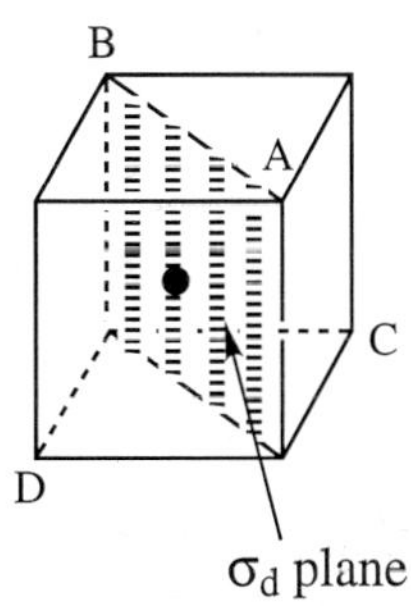

Figure 1.18: Illustration of various symmetry operations of tetrahedral AB_4 molecule within a cube .

1.5.4 AB_6 Type of Molecule (Octahedral Geometry)

Here the octahedral geometry has been represented in three different ways (I, II and III) for easy visualization and understanding. In octahedral geometry, the symmetry operations present E, $8C_3$, $6C_2'$, $6C_4$, $3C_2(=C_4^2)$, i, $6S_4$, $8S_6$, $3\sigma_h$, $6\sigma_d$.

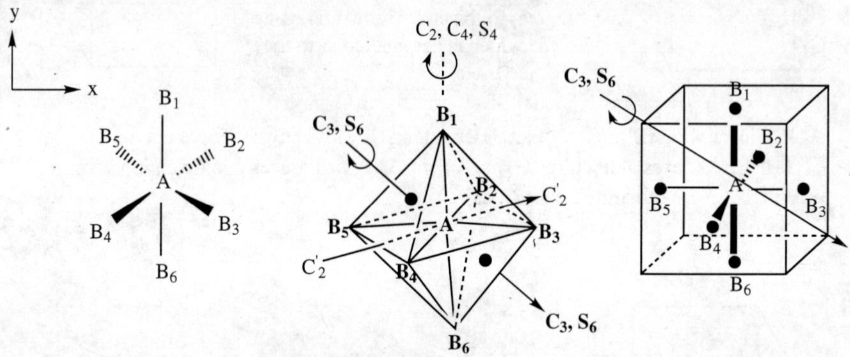

(I) (II) (III)

Figure 1.19: Different representations of octahedral, AB_6 molecule along with its different symmetry elements .

Some important features of the Octahedron:

- There are 6 vertices (3 opposite pairs of vertices; *e.g*, B_1 B_6; B_2 B_4, *etc.*).
- There are 8 trigonal planes (4 pairs of opposite trigonal planes; *e.g.*, $B_1B_4B_5$ and $B_2B_3B_6$ are two opposite trigonal planes. Similarly, other 3 pairs or 6 trigonal planes are there).
- There are 12 edges or 6 pairs of opposite edges (line between any two B's). Line connecting B_3B_4 is an edge and B_2B_5 edge is opposite to the B_3B_4 edge, similarly B_1B_4 is opposite edge of B_2B_6. Another four pairs are there.
- Point, *A* is the center of the O_h in space.
- **C_3 axis:** Although a little bit tricky to imagine a C_3 axis, but if we take the above pictorial representation (*II*), then it is not so difficult to find this axis. Here, we have to find pair of opposite trigonal planes which have already been demonstrated. If a line is considered through the center of such two opposite planes, that can be designated as C_3 axis. For example, imagine a line passing through the center of two planes $B_1B_4B_5$ and $B_2B_3B_6$ as well as through central point A. Similarly, we have other three pairs of such planes and hence corresponding another three C_3 axes can be found. From representation (*III*), C_3 is considered to passing through two opposite diagonal corners of an octahedron. Each of these four C_3 axes produces two operations (C_3^1 and C_3^2). Hence, there will be total 8 such operations; four C_3 operations and four C_3^2 operations.
- **C_4 axis:** C_4 is passing through two opposite vertices (e.g., Through B_1 and B_6 or B_2 and B_4 or B_3 and B_5). Three such pairs are there and hence there will be such three C_4 axes. Each of these C_4 axes will generate C_4^1, C_4^2 and

C_4^3 operations. Hence, total **nine operations** will be generated from C_4 axes. But careful inspection shows that among nine operations, three operations (C_4^2 operations) can be replaced by three C_2 operations and ultimately C_4 axis will produce **total six operations (3C_4 operations and 3C_4^3 operations)**.

- **C_2 axis:** As discussed above, three C_2 axes are there which are collinear with C_4 axis and they are basically C_4^2 operations through each pair of opposite vertices.
- **C_2' axis:** Apart from those **three C_2** axes which passes through opposite vertices, another type of C_2 axes are there which passes through opposite edges. We can illustrate one such C_2 axis if we consider a line which passes through the center of B_2, B_3 edge and B_4, B_5 edge and also through point A. There are such six pairs of opposite edges and each will give one such C_2' axis. **Total six such C_2' axis.**
- Inversion point is at A.
- **S_4 axis:** S_4 axis is collinear with each C_4. There are three S_4 axes through each pair of opposite vertices and each S_4 axis produces two operations (So total six operations; **3S_4^1** and **3S_4^3 operations**).
- **S_6 axis:** Four S_6 axes are collinear with four C_3 axes. Each S_6 axis generates two symmetry operations. Hence, total eight operations are there; **4S_6^1** and **4S_6^5 operations**. Other S_6 operations can be substituted by C_3 or C_3^2 or *i* or E operations.
- **σ_h planes:** If we have found properly three C_4 axes, then it is now a very trivial job to find three σ_h planes. As C_4 axis is the principal axis for O_h geometry, hence the three planes perpendicular to the corresponding three C_4 axes will be total **three σ_h planes**.
- **σ_d planes:** σ_d is the plane bisecting two C_2 axes. Let us consider a plane containing B_1 and B_6 vertices through the central point A, bisecting the angle $\angle B_3AB_4$ (through yz plane). Now if we go through the picture (I) very carefully, we can see that a similar plane can be generated considering the same two vertices but bisecting the angle $\angle B_2AB_3$ (through xy plane). Hence, we can see that each pair of vertices can produce two σ_d planes on two mutually perpendicular planes (In this case, xy and yz planes) which can bisect two mutually perpendicular C_2's, passing through B_2AB_4 and B_3AB_5. Since, there are such three pairs of opposite vertices, hence **total six such σ_d planes** will be there. If we inspect carefully and recall the six C_2 axes, it is found that each σ_d plane indeed bisect two C_2 axes.

Few important representative operations, which are a little bit tricky in nature, are shown below in Figure 1.20. Remember that here all the rotations are considered clockwise.

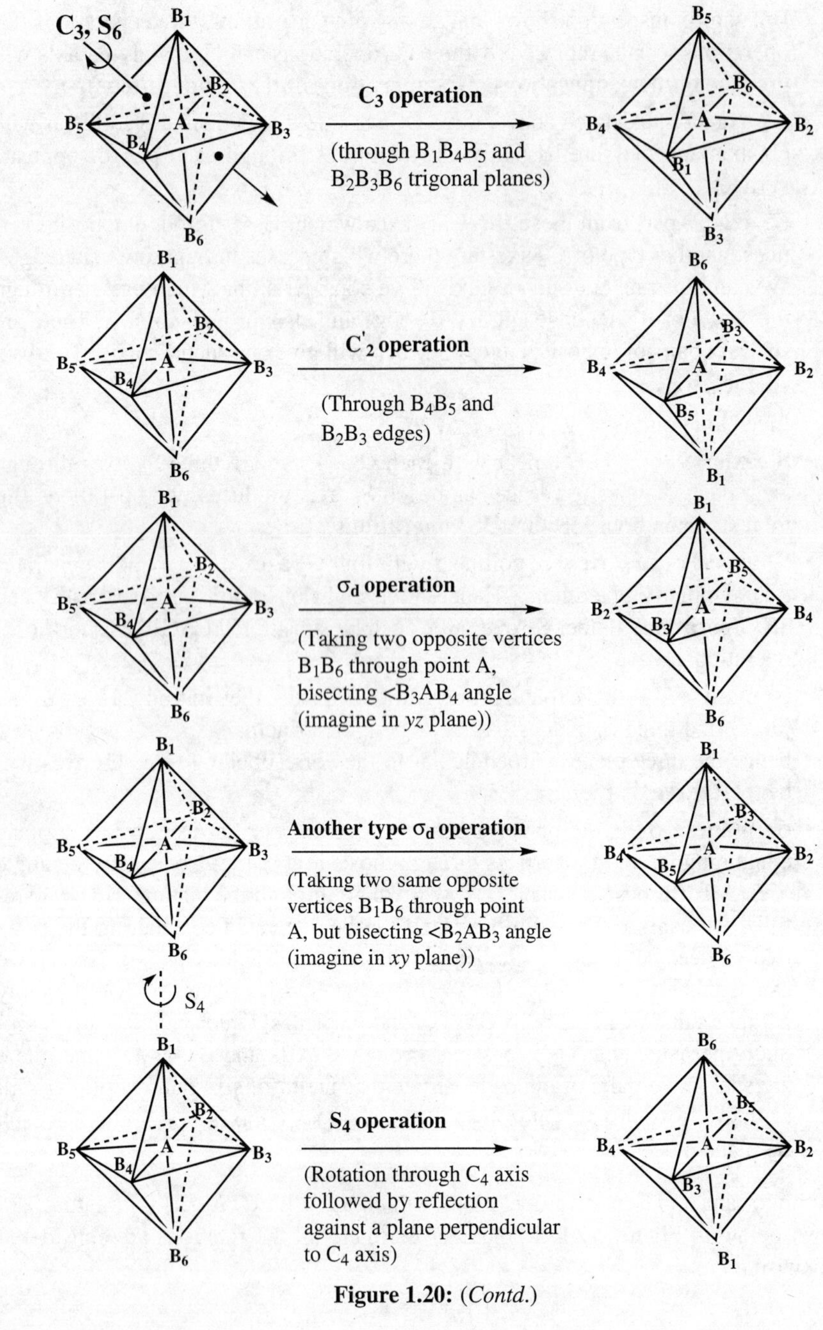

Figure 1.20: (*Contd.*)

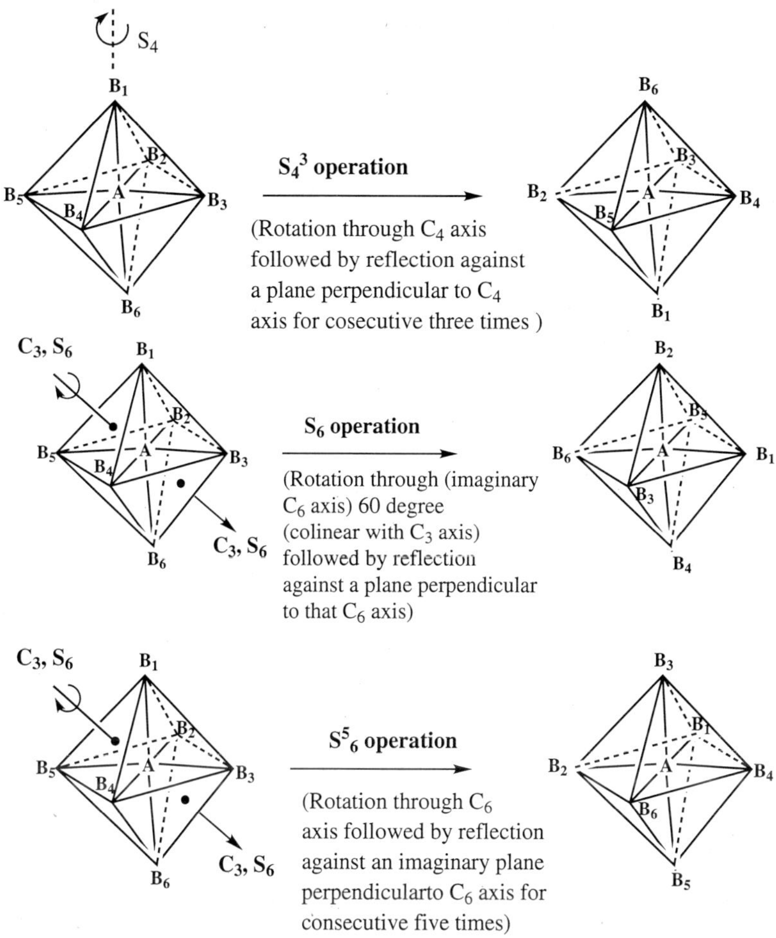

Figure 1.20: Illustration of various symmetry operations of octahedral AB_6 type of molecule.

EXERCISES

1.1 What is the difference between *symmetry element* and *symmetry operation*?

1.2 What are the symmetry elements present in (i) PCl_3 (ii) C_2HF (iii) C_2H_2 (iv) SO_3 (v) BrF_5 (vi) Ferrocene (*staggered*, *eclipsed*) (vii) Ethane (*staggered*, *eclipsed*, *skewed*) (viii) H_2O_2 (ix) *cis* and *trans*-dichloroethylene (x) CHClBrF?

1.3 Construct multiplication table for the corresponding symmetry operations of molecules with C_{4v} and D_{6h} point groups.

1.4 Set up the multiplication table for the symmetry operations of *trans*-2 butene molecule.

1.5 Show the commutative property among different symmetry operations of (i) NH_3 molecule (ii) BCl_3 molecule.

1.6 Which improper axis of symmetry is present in BCl_3 molecule? Find the different equivalent symmetry operations for its possible improper axis of symmetry operations.

1.7 Predict the direction of dipole moment of NH_3 molecule with the help of its symmetry aspects.

1.8 Make a comparative discussion of symmetry elements of ClF_3 and BF_3.

1.9 List the symmetry operations for the following molecules:
(i) Toluene (considering methyl group as a sphere) (ii) Toluene (considering methyl group hydrogens as tetrahedrally disposed) (iii) *o*, *m* and *p*-dichlorobenzene (iv) *p*-bromochlorobenzene (v) Cyclohexane (boat and chair) (vi) A tree.

Chapter 2

Group, Subgroup, Point Group and Conjugate Class

2.1 Group

So far we have seen that for different molecules there are different combinations of five symmetry elements (*i.e.*, E, C, S, i and σ) present. For example, water, pyridine, m-xylene, phenanthrene molecules have the same set of symmetry elements, namely E, C_2, σ_v, and σ'_v. Similarly, BCl_3 and $C_3H_3^-$ have the same set of symmetry elements (E, C_3, C_2, σ_h, σ'_v and S_3). Therefore, it is convenient to classify all such molecules by a single symbol which summarizes all the information regarding their symmetry.

The first set of molecules is designated by C_{2v} and the second set of molecules are designated by D_{3h}. These are known as the *Schoenflies symbol* or the notation of the *point group*. This can be summarized as follows:

1. The full set of symmetry operations that a particular molecule contains is called a *point group*.
2. There are specific symbols for each of the point groups.
3. These symbols are often used to describe the symmetry of a molecule.
4. Rather saying water is bent, from now we will say water has C_{2v} symmetry.
5. Point group symbol consists of **three** parts:
   (a) The number at the subscript indicates the order of principal axis (*i.e.*, highest order axis). This is conventionally taken as vertical axis.
   (b) The small letters (h, v) in subscript indicate, respectively, the presence and absence of horizontal plane.
   (c) The capital letters (C or D) indicate, respectively, the absence and presence of n number of C_2 axes at right angles to the principal C_n axis. *e.g.* in Benzene (D_{6h}), principal axis is C_6 and there are six C_2 axes perpendicular to the principal axis.

It is already mentioned that all these symbols represent the point group of a molecule. The word *point* is used because all the symmetry elements of a molecule always pass through one common point (which sometimes passes through a line or a plane or through space) *e.g.*, if we take the example of benzene, all the possible symmetry elements present, passes through a common point which is situated at the center of benzene ring in space. Now, the term GROUP has a precise mathematical meaning and the set of symmetry operation of a molecule constitutes a mathematical group.

A group generally obeys four rules:

1. **Closer property**: If P and Q belongs to group G, and binary combination (say, multiplication) of P and Q, such that, $P \times Q = R$, then R also belongs to the same group, G.
2. **Identity element**: There must be a unique member in every group (identity element) apparently which has no role. Binary operation between this unique element with any other member of a group results that particular member.
3. **Associative property**: If A, B and C are the three members of group, then combination among these elements must be associative unless the order of A, B, C changes, *i.e.*, $(A \times B) \times C = A \times (B \times C)$.
4. **Inverse element**: Every member of a group must have an inverse which is also a member of the group, *i.e.*, $A \times A^{-1}$ = E, the identity. If A is a member of a group then A^{-1} must also be a member of that group. Sometimes a member is inverse of itself or it may be any other member of that group.

Let us take a group, G which consists of the four elements $\{1, -1, i, -i\}$ and check whether these elements follow the above mentioned four rules under the multiplication operation.

1. It is found that $1 \times 1 = 1$, $i \times i = -1$, $1 \times i = i$, $(-i) \times (-i) = -1$, $i \times (-i) = 1$

   Since, the products also belong to the group, G Hence, it follows the *closer property*.
2. Here, the *identity element* is 1 since, after binary combination with other members of the group under multiplication, apparently has no effect.
3. It is also found that, $[(1 \times (-1)) \times i] = [1 \times ((-1) \times i)]$ *i.e.*, independent of the sequence of operation. Hence, here *assosciative property* is valid.
4. $1 \times 1 = 1$; Hence, here 1 is found to be its own inverse, since the result of the binary product is *identity element*; Similarly, $i \times (-i) = 1$ (produces the *identity element*); Hence, $-i$ is the inverse of i and vice-versa; Also, $(-1) \times (-1) = 1$; and -1 is the inverse of -1 itself. Therefore, it is evident that each element has its own *inverse element* and is present as a member of the group of elements as well.

Since, the four elements of the group, G follows the basic four rules of a GROUP hence, they constitute a GROUP. The members of a group or which are called *elements*, may be constituted with some numbers or some matrices or may be symmetry operators *etc*. Right now,without bothering about the symbol, let us take an example of molecule which belongs to C_{2v} point group (*e.g.*, H_2O molecule). This group has four symmetry operations, E, C_2, σ and σ' are taken as members of the group named, C_{2v}.

At this point, we need to recall the multiplication table for C_{2v} point group (Table 1.2). From the table, the rules (i), (ii) and (iv) can easily be verified for a group. Here, the identity element is E as it does not change any element after binary operation. The inverse of any element can easily be found by simply searching the specific element with which after binary operation, produces the identity element.

For rule (iii), the sequence of operations are performed $(C_2 \times \sigma) \times \sigma' = \sigma' \times \sigma' = E$; Again, $C_2 \times (\sigma \times \sigma') = C_2 \times C_2 = E$. Hence, it holds the associative property. This rule can also be verified taking other sets of elements as well.

Another exercise can be tried by taking the example of C_{3v} point group. Here, again, the corresponding multiplication table (Table 1.3) can be recalled and verified as to the four essential criteria for being a group. Here we have the element E (Identity element) which after combination with other elements does not change the character of the other participating element. After binary operation between any two elements, the resultant element is always within these six elements (*Closer property*). From the multiplication table, it is clearly evident that C_3 element is the inverse of C_3^2 element and vice versa whereas σ's are the inverse of itself. Hence, it is found that every element has their own inverse. The associative property of elements can be verified as: $(\sigma' \times \sigma'') \times C_3 = C_3 \times C_3 = C_3^2$ and $\sigma' \times (\sigma'' \times C_3) = \sigma' \times \sigma = C_3^2$.

Since, $(\sigma' \times \sigma'') \times C_3 = \sigma' \times (\sigma'' \times C_3)$; Hence, associative property is well maintained. So, it can be concluded unanimously that the symmetry elements of C_{3v} point group form a symmetry group.

2.2 Subgroup

As already been mentioned that one of the main objectives to study chemical applications of group theory is to understand the change in molecular properties with structural changes. It is found that molecular properties changes with substitution of atoms or rearrangement of atoms in a molecule. During this atomic substitution or rearrangement, there is always a change in the order (total number of symmetry operations) of a point group. If after substitution or rearrangement, the order of the new molecule or point group is less, it is said that descent in symmetry has occurred and if there is an increase in order of new point group, it is understood that ascent in symmetry has occurred. It is found that during ascending of point group, degeneracy increases and during descending of point group, lifting of degeneracy occurs. When there is descent in symmetry in new point group, in many cases, it is found to be a subgroup of the old (higher order) point group. A *subgroup* of a group is a part of a group which has all the four group properties under the same binary operation. For example, the group $G = \{1, -1, i, -i\}$ where the binary is multiplication. Here the subset $\{1, -1\}$ exhibits all the four group properties under multiplication operation, as can be readily verified. Thus, $\{1, -1\}$ is a subgroup of group G.

Let us take the example of C_{3v} point group (order 6), where the symmetry operations are $E, C_3, C_3^2, \sigma_v, \sigma_v', \sigma_v''$. The symmetry operations can be grouped as follows: $\{E\}, \{E, \sigma_v\}, \{E, \sigma_v'\}, \{E, \sigma_v''\}, \{E, C_3, C_3^2\}$.

According to the presence of members (or symmetry operations) in the above mentioned groups, they can be named as C_1, C_s, C_s, C_s and C_3 respectively. Since, these groups are generated from the parent group, C_{3v}, hence, these three groups, C_1, C_s and C_3 are called as subgroup with order 1, 2 and 3 respectively.

In an another example of D_{2h} point group (order 8), the symmetry operations are $E, C_2(z), C_2(x), C_2(y), i, \sigma_h, \sigma_{v_1}$ and σ_{v_2}.

Here possible subgroups are:
$[E], [E, C_2(z)], [E, \sigma_h], [E, i], [E, C_2(z), \sigma_h, i], [E, C_2(z), \sigma_{v_1}, \sigma_{v_2}]$ and $[E, C_2(z), C_2(x), C_2(y)]$ or these can be named according to their point group nomenclature as, C_1, C_2, C_s, C_i, C_{2h}, C_{2v} and D_2 with order (h)= 1, 2, 2, 2, 4, 4 and 4, respectively. These smaller sets, containing symmetry operations, are groups on their own (since, obey the four basic requirements of a group) and can be considered as subgroups of the larger parent group from which the elements were scrapped. In general, if g is the order of a subgroup of a group whose order is h, then $h/g = n$, where n is an integer. In other words, the order of any subgroup must be an integer divisor of the order of the parent group. However it is not a requirement that subgroups of all the allowed orders must exist. For C_{2v} (where h=4), only subgroups with orders g=1 and g=2 are possible, as integer 4 has allowed divisors, 1 and 2 only. In this simple case, subgroups of both the orders exist. Group-subgroup relation can be demonstrated by taking few molecules as example, shown in Figure 2.1.

Let us take an octahedral (O_h) molecule MA_6 (structure *I*, Figure 2.1), where there are 48 symmetry operations (*i.e.*, Order, h= 48). Now, if atoms are substituted continuously, as shown in figure, different point group symmetries (D_{4h}, C_{4v}, C_{2v} and C_{3v}) are reached. In Table 2.1, order of all the point groups along with their full list of symmetry operations are provided. Hence, from the criteria of being a subgroup of a parent group it is seen that the structure *II*, *III*, *IV* and *V* are subgroups of structure *I* although *V* belongs to different successive series. It is also found that in every case of subgroup, they lose few key symmetry elements down the descending series.

Table 2.1: Examples of group and subgroups along with their list of operations.

Structure No.	Point Group	Order(h)	List of operations
I	O_h	48	$E, 8C_3, 6C_2, 6C_4, 3C_2, i, 6S_4, 8S_6, 3\sigma_h, 6\sigma_d$
II	D_{4h}	16	$E, 2C_4, C_2, 2'_2, 2C''_2, i, 2S_4, \sigma_h, 2\sigma_v, 2\sigma_d$
III	C_{4v}	8	$E, 2C_4, C_2, 2\sigma_v, 2\sigma_d$
V	C_{3v}	6	$E, 2C_3, 3\sigma_v$

It should be noted that if the order of any subgroup is not an integer divisor of the order of the parent group, then without inspection of their list of symmetry, it can be concluded that they do not hold the group-subgroup relation. For example, MA_5 can have square pyramidal ($C_{4v}, h = 8$) or trigonal bipyramidal ($D_{3h}, h = 12$) geometry. But, as they do not follow the criteria ($h/g = n$), hence they do not hold group-subgroup relation.

Now, let us take few examples to illustrate how nature of point group of a molecule changes with substitution of atoms.

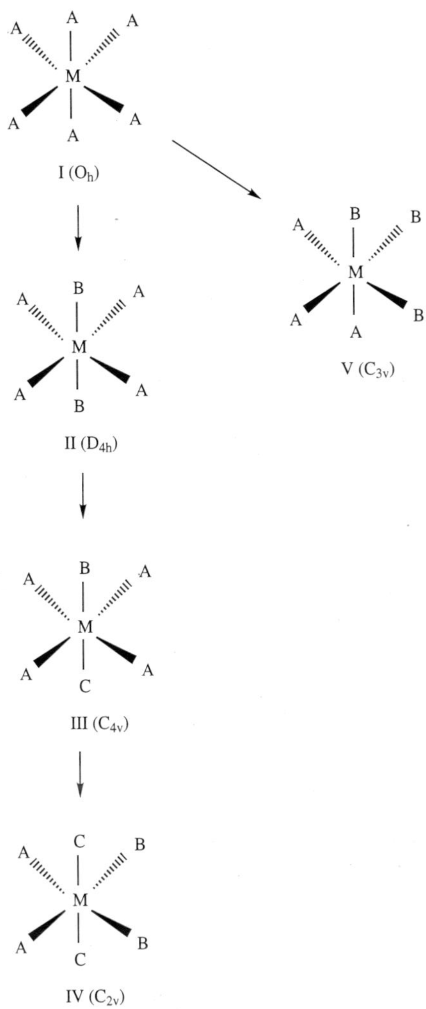

Figure 2.1: Illustration of group-subgroup relation.

2.3 Cyclic Group

These are also known as single-axis rotational groups. A group of the type $(x, x^2, x^3, x^4, \ldots x^n = E)$ is called a *cyclic group* with generator x. For example, C_n family consists of the operations generated by a C_n rotation carried out successively n times, *i.e.*, $C_n^1, C_n^2, C_n^3 \ldots, C_n^n = E$. For example, the group of cube roots of unity, $G = (\omega, \omega^2, 1)$ is cyclic group of order 3 with generator ω (where $\omega^3 = 1$). In the point group, $C_{3v} = (E, C_3, C_3^2, \sigma_v, \sigma_v', \sigma_v'')$, we have the subset, (E, C_3, C_3^2) as a cyclic subgroup of C_{3v} of order 3 with generator C_3. Similarly (E, σ_v) is a cyclic subgroup of C_{3v} with order 2 and generator σ_v. Some interesting observations are there for cyclic groups in their multiplication table . For example, C_4 point group consists of the operations,

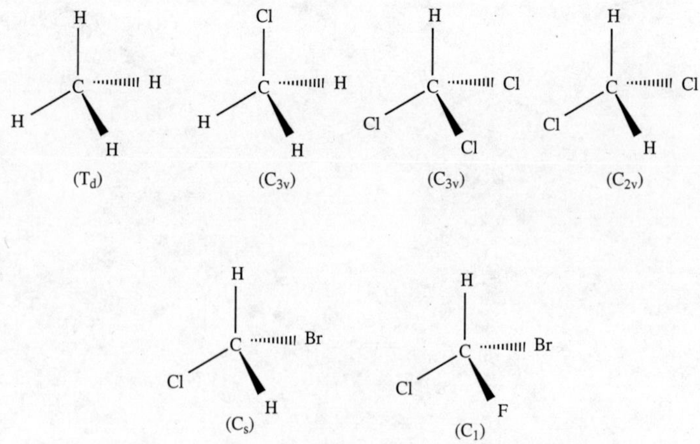

Figure 2.2: Effect of substitution on nature of point group.

$E, C_4, C_4^2 (= C_2)$ and C_4^3 and the multiplication table for these operations is presented in Table 2.2.

Table 2.2: Multiplication table of C_4 point group

C_4	E	C_4	C_2	C_4^3
E	E	C_4	C_2	C_4^3
C_4	C_4	C_2	C_4^3	E
C_2	C_2	C_4^3	E	C_4
C_4^3	C_4^3	E	C_4	C_2

It is found that the product in every row or the elements of every row follow same cyclic sequential order as shown in Figure 2.3. Because of this trend of the elements, if we look at the table very carefully, it is found that a particular element appears right-to-left in diagonal positions in the table. With this information, it becomes easy to construct the multiplication table of any cyclic group.

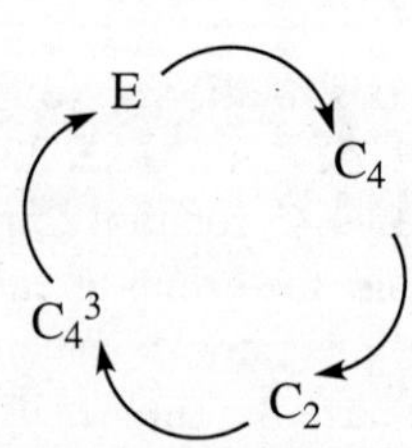

Figure 2.3: Scheme for cyclic sequential order.

2.4 Conjugate Class and their Properties

2.4.1 Similarity Transformation

There is a well-defined mathematical procedure to find out which operations of point group are 'similar' to each other, which is called *similarity transformation*. A similarity transformation in group theory is a transformation that maps an symmetry operation to a conjugate symmetry operation within a group. In symmetry operations, a similarity transformation represents a change of basis that preserves fundamental properties of the system.

Let G = $\{E, R_2, R_3, ...R_i, R_j, R_p....\}$ be a group. If $R_p^{-1} \times R_i \times R_p = R_j$, then R_i and R_j are said to be *similarity transform* of one another or *conjugate* to each other. Here, R_i is the target element on which similarity transformation is executed and R_j is the resultant element. The effect of R_p on some object is nullified by the inverse operation R_p^{-1}. Hence in the composite operation $R_p^{-1} \times R_i \times R_p$ we expect the result to be either R_i itself or some operation whose effect is equivalent or very similar to that of R_i.

Here, the symmetry multiplication table of NH_3 (Table 2.3) will be of great use to find the conjugate relationship among the elements of the point group.

Table 2.3: Multiplication table of C_{3v} point group

C_{3v}	**E**	C_3	C_3^2	σ	σ'	σ''
E	E	C_3	C_3^2	σ	σ'	σ''
C_3	C_3	C_3^2	E	σ''	σ	σ'
C_3^2	C_3^2	E	C_3	σ'	σ''	σ
σ	σ	σ'	σ''	E	C_3	C_3^2
σ'	σ'	σ''	σ	C_3^2	E	C_3
σ''	σ''	σ	σ'	C_3	C_3^2	E

Let us consider C_3 as the target element. The possible similarity transformations on C_3 are:

$(E)^{-1}C_3E = C_3$

$(C_3)^{-1}C_3C_3 = C_3^2C_3C_3 = C_3$

$(C_3^2)^{-1}C_3C_3^2 = C_3C_3C_3^2 = C_3^4 = C_3$

$(\sigma)^{-1}C_3\sigma = \sigma C_3\sigma = \sigma\sigma'' = C_3^2$

$(\sigma')^{-1}C_3\sigma' = \sigma' C_3\sigma' = \sigma'\sigma = C_3^2$

$(\sigma'')^{-1}C_3\sigma'' = \sigma'' C_3\sigma'' = \sigma''\sigma' = C_3^2$

Hence, it is evident that similarity transformation on C_3 always results either C_3 or C_3^2 and consequently it can be concluded that these two symmetry operations are in the same class. Similarly, it would be observed that applying similarity transformation on C_3^2, produces either C_3 or C_3^2 which prompt us to conclude that C_3 and C_3^2 are conjugate to one another or they are in the same class. The number of symmetry operations in a conjugacy class is called the class size.

Subsequently, it can also be found that $\sigma, \sigma', \sigma''$ are conjugate to one another when those three symmetry operations is taken as target element. It should be noted that E is always a class by itself as E commute with all other elements. Therefore, in C_{3v} point group there are three classes; first one with a single member, E, second one with two members, C_3 and C_3^2 and the third one comprises of the three vertical planes, σ, σ' and σ'', hence operations as class wise: $[E, (C_3, C_3^2)$ and $(\sigma, \sigma', \sigma'')]$.

At this point it should be mentioned that in H_2O (C_{2v}), there are four classes namely, $[E, C_2, \sigma, \sigma']$. In BCl_3 (D_{3h}), there are six classes:

$$[(E), (C_3, C_3^2), (C_2, C_2', C_2''), (\sigma_h), (S_3, S_3^5), (\sigma, \sigma', \sigma'')]$$

2.4.2 Few Interesting Salient Features of Conjugate Classes

1. E, σ_h and i are such unique elements in molecular point group which form class by itself.
2. Principal axis, (C_n) and its inverse (C_n^{-1}) belong to the same class, if either σ_v or such C_2 axis which is perpendicular to C_n axis, is present in the point group.
3. S_n and its inverse S_n^{-1} belong to the same class, if either σ_v or such C_2 axis, which is perpendicular to S_n axis, is present in the point group.
4. Except principal axis, other axes will be in same class, if atoms or groups present in one axis can be moved to other axis by means of some other symmetry element, present in that point group. For example, in C_{3v} point group, C_3 and C_3^2 operations are in same class as only nitrogen atom in ammonia is involved here in both cases. In $D_{4h}(PtCl_4)$, the axes passing through only Pt atom are in same class, whereas axes passing through Pt and other two C_l atoms belong to other class.
5. Two symmetry planes σ_v and σ_v' will be in same class, if atoms or groups present in one plane can be moved to other plane by means of some other symmetry element present in that point group. In C_{2v} point group (Water molecule), one plane contains only O-atom and other plane contains all the three atoms of water molecule. A single operation cannot fulfill the criteria stated above. But in case C_{3v} (Ammonia molecule), all the three planes contain N-atom and one H-atom. One C_3 operation can change the atoms from plane to other. Therefore, the three planes are in the same class.

2.4.3 Important Properties of Conjugate Classes

1. *The identity element, E forms a class by itself.*

   Proof: For any element X of a group under consideration,
   $X^{-1}EX = X^{-1}X = E$
   Since, this is true for any X, hence, E is conjugate to itself only.
2. *If an element X of group commutes with all other elements of the group, then X forms a class by itself.*

   Proof: Let A be a member of a group G.
   If, $AX = XA$, then $A^{-1}XA = A^{-1}AX = X$

Since, A may be any member of G, the above result is true for all A in G. Therefore, all similarity transformations on X will give the same result X. Thus, X forms a class itself. Note: A group is called *abelian* if each element of a group commutes with every other elements of a group. Examples are C_{2v}, D_2, C_n *etc*. In such a group every element forms a class by itself.

3. *No element of a group can belong simultaneously to two different classes.*

   Proof: If possible, let C_1 and C_2 be two different classes of a group G and let P be an element common to them.
   Let, $X \in C_1$ and $Y \in C_2$
   X and Y should not be conjugate to one another since, they belong to different classes.
   Again, let $P \in C_1$ and $X \in C_1$
   Hence, P and X are conjugate, *i.e.*, $A^{-1}PA = X$ for some A in G.
   Now, left multiplying both sides by A and right multiplying by A^{-1} we get,
   $AA^{-1}PAA^{-1} = AXA^{-1}$ Or, $P = AXA^{-1}$
   Similarly, since $P \in C_2$ and $Y \in C_2$, we can write
   $P = BYB^{-1}$ for some element, B in group, G
   Then, $AXA^{-1} = BYB^{-1}$
   Now left multiplying both sides by A^{-1} and right multiplying by A we have,
   $X = A^{-1}BYB^{-1}A = Z^{-1}YZ$ (where $Z = B^{-1}A$ such that $Z^{-1} = A^{-1}B$)
   Owing to closure property and existence of inverse of each group element, both Z and Z^{-1} belong to G.
   Hence according to the above relation, X and Y are conjugate to one another, which is contrary to the starting hypothesis or assumption.
   Hence, P can not be common to C_1 and C_2. Thus, it can be concluded that no element of a group can belong simultaneously to two different classes.

4. *If two elements of a group are conjugate to a third element, then they are themselves conjugate.*

   Proof: Let A, B and C are three elements of a group, G such that A is conjugate to C and also B is conjugate to C. We have to prove that A and B are conjugate to one another as well. Since, A is conjugate to C, hence, $X^{-1}AX = C$ for some X $\in$ G.
   Similarly, since B is conjugate to C, hence, $Y^{-1}BY = C$ for some other element Y $\in$ G.
   Equating both the relations, we can write,
   $X^{-1}AX = Y^{-1}BY$
   Or, $A = XY^{-1}BYX^{-1} = (YX^{-1})^{-1}B(YX^{-1})$.
   Since, X and Y belong to G, YX^{-1} also belongs to G. Hence A and B are conjugate to one another.

5. *If A and B are in the same class and $A^h = E$, then $B^h = E$ (h is an integer, called period of an element)*

   Proof: Let us consider that A and B are conjugate.

   Therefore, it can be written, $X^{-1}AX = B$ for some $X \in G$
   Since, $A^h = E$

Hence, $A.A.A......A(h \text{ factors}) = E$
Or, $AXX^{-1}AXX^{-1}AXX^{-1}A.....XX^{-1}A = E$
Or, $AXB.B.B.....B.X^{-1}A = E$
Or, $AXBh^{-2}X^{-1}A = E$
Or, $X^{-1}AXB^{h-2}X^{-1}AX = X^{-1}EX$
Or, $BB^{h-2}B = E$
Therefore, $B^h = E$ (Proved)

Note:

1. h is said to be the period of an element, A of a group if $A^h = E$. This theorem says that elements of any class of a group have the same period.

   For example, in C_{3v}, the classes are $(E), (C_3, C_3^2), (\sigma_v, \sigma_v', \sigma_v'')$. Here $C_3^3 = E$, *i.e.*, C_3 has period 3. Similarly, it is seen that$(C_3^2)^3 = E$, *i.e.*, C_3^2 has period 3 as well.

2. The converse of the above is not always true. For example in C_{2v} point group, $C_2^2 = E$ and $\sigma_v = E$ but C_2 and σ_v are not in the same class.

Try: A multiplication table (Table 2.4) under some binary operation Δ is presented. Verify that the set of elements, (A, B, C, D) forms a group under the binary operation Δ. Find the results of $A\Delta B\Delta C^{-1}$, $B^{-1}\Delta D\Delta B$ and $A^{-1}\Delta C\Delta A$ as well. Also show that the group is *Abelian*.

Table 2.4: Multiplication table under operation, Δ.

Δ	**A**	**B**	**C**	**D**
A	B	A	D	C
B	A	B	C	D
C	D	C	B	A
D	C	D	A	B

- From the table it is very much clear that identity element is B since, it does not change the participating element after binary operation.
- Every element is found to be its own inverse, as every element after binary operation with itself results identity element, B.
- Since, from the multiplication table, it is found that binary combinations between any two elements always produce only those four elements, hence obviously closer property is nicely followed.
- Since, $(A\Delta C)\Delta B = D\Delta B = C$ and $A\Delta(C\Delta B) = A\Delta D = C$
  Hence, $(A\Delta C)\Delta B = A\Delta(C\Delta B)$; Associative property holds.

$$ABC^{-1} = ABC = AC = D$$

$$B^{-1}DB = BDB = BD = D$$

$$A^{-1}CA = ACA = AD = C$$

- To prove the group as abelian, we have to show that every element commutes with every other elements.

$$AB = A; BA = A; \text{ Hence, } A, B \text{ commute.}$$

$$AC = D; CA = D; \text{ Hence, } A, D \text{ commute.}$$

$$AD = C; DA = C; \text{ Hence, } A, D \text{ commute.}$$

Hence, A commutes with B, C and D. Similarly, it can be shown for other elements as well and hence this set of elements can be considered as abelian group.

Table 2.5: Few common examples of molecules along with their point groups.

Point Group	**Symmetry elements**	**Examples**
C_1	E	CHIBrF
C_i	E, i	*staggered* CHClBr-CHClBr
C_s	E, σ_h	HOCl, 1-Methylimidazole
C_n	E, C_n	$H_2O_2, skewed\ CH_3 - CCl_3$
C_{2h}	E, C_n, σ_h	*trans-dichloroethylene*
C_{2v}	$E, C_2, \sigma_v, \sigma_v'$	Water, *Cis*-Dichloroethylene, Dichloromethane, *Cis*-$[PtCl_2Br_2]$, Chlorobenzene, Toluene (taking methyl group as one big atom), *o*-Dichlorobenzene, *m*-Dichlorobenzene, *p*-Bromochlorobenzene, 1,2,3-trichlorobenzene, Pyridine, Cyclohexane (boat)
C_{3v}	$E, C_3, 3\sigma_v$	Ammonia, Nitrogen trifluoride, Chloroform, Trichloromethane, Toluene
C_{4v}	$E, C_4, C_2, 4\sigma_v$	Square pyramidal molecule
D_n	E, C_n, nC_2	*skewed* Ethane
D_{2h}	$E, 3C_2, 2\sigma_v, i, \sigma_h$	Ethylene, *tran-* $[PtCl_2Br_2]$, *p*-Dichlorobenzene, Naphthalene, Biphenyl (planar), Pyrene
D_{3h}	$E, C_3, 3C_2, 3\sigma_v, \sigma_h, S_3$	*eclipsed* Ethane, 1,3,5-trichlorobenzene, BCl_3, Cyclopropane
D_{4h}	$E, C_4, 4C_2, 4\sigma_v, i, \sigma_h, S_4$	$[PtCl_4]^{-2}$
D_{6h}	$E, C_6, C_3, C_2, 6C_2, 6\sigma_v, i, \sigma_h, S_6$	Benzene
D_{3d}	$E, C_3, C_2, 3C_2, 3\sigma_d, S_6$	*staggered* Ethane, Cyclohexane (chair)
$D_{\infty h}$	$E, C_\infty, \infty C_2, i, \sigma_h, \infty\sigma_v$	Acetylene, H_2, CO_2
$C_{\infty v}$	$E, C_\infty, \infty\sigma_v$	HCl, SCO

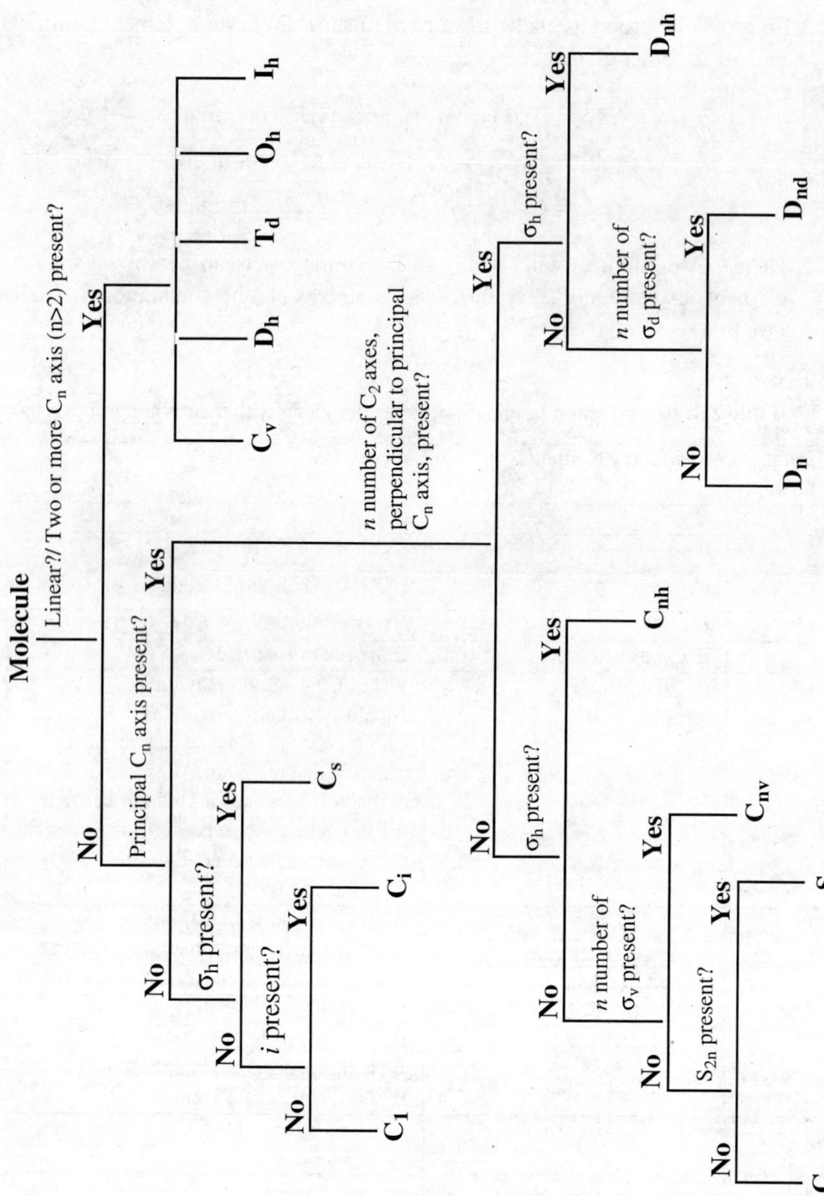

Figure 2.4: Interactive flow chart for systematic determination of point group of a molecule.

EXERCISES

2.1 Verify the associative property of C_{3v} point group.

2.2 Discuss, how point group of NH_3 molecule changes as its H-atoms are replaced by Cl-atoms systematically.

2.3 Show that every symmetry operation of H_2O molecule belongs to different classes.

2.4 Establish the different classes of NH_3 molecule.

2.5 Comment on the *abelian* group properties of H_2O and NH_3 molecules.

2.6 What are the point groups of the following molecules:
(i) dichlorobenzene (*ortho, meta, para*) (ii) Chloroethylene (iii) *cis* and *trans*-dichloroethylene (iv) 1,2,4-trichlorobenzene (v) 1,3,5-trichlorobenzene (vi) BCl_3 (vii) BBrClF (viii) $B(OH)_3$.

Chapter 3

Representation of Symmetry Operations: Reducible and Irreducible Representation

3.1 Representation of Symmetry Operations

So far we have not tried to represent symmetry operations in a quantitative manner. They can be represented in a quantitative manner as well, when Symmetric or Antisymmetric orbitals, Unit vectors, Rotational vectors, Cartesian coordinates, Position vectors, Orbital wave functions, Mode of vibrations *etc.*, are taken as basis and consequent symmetry operations are performed on them. *Basis is a set of wave functions, orbitals, coordinates or vectors etc., on which symmetry operations are carried out to construct their corresponding matrix representation.* It would be clear if we discuss this with some examples.

3.1.1 P_x-Orbital as Basis

The effects of different symmetry operations on $\mathbf{p_x}$**-orbital** has been shown in Figure 3.1.

The convention is that if there is no change in the direction of the orbital lobes upon execution of a particular symmetry operation, then it contributes +1, and if the lobe changes direction, the contribution is taken as –1 and these numbers are called character (Γ) corresponding to each operation. In the first case, the p_x orbital is said to be **symmetric** with respect to those operations, and in the second case, the orbital or function is said to be **antisymmetric** with respect to those operations. Here, we can see that the p_x orbital is symmetric with respect to E, $C_2(x)$, σ_{xy} and σ_{yz} operations and antisymmetric with respect to $C_2(y)$, $C_2(z)$, σ_{yz} and i operations.

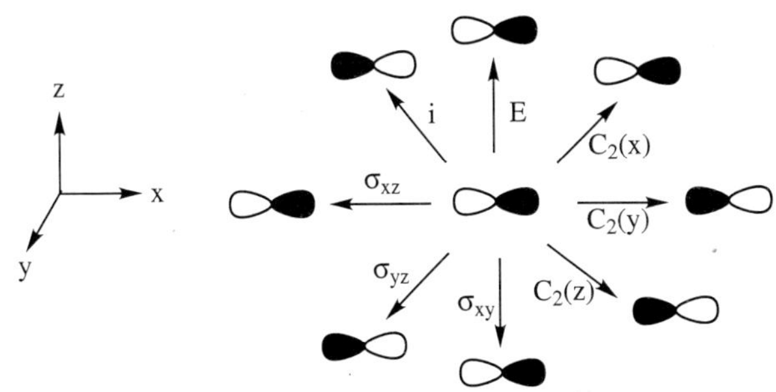

Figure 3.1: Operation of various symmetry operations on p_x orbital

The effect of various symmetry operations on p_x orbital can be summarized as follows:

p_x	E	$C_2(x)$	$C_2(y)$	$C_2(z)$	σ_{xy}	σ_{yz}	σ_{xz}	i
Γ_1	+1	+1	–1	–1	+1	–1	+1	–1

Similarly, characters for p_y and p_z orbitals can be constructed and tabulated:

Basis	Representation	E	$C_2(x)$	$C_2(y)$	$C_2(z)$	σ_{xy}	σ_{yz}	σ_{xz}	i
p_y	Γ_2	+1	–1	+1	–1	+1	+1	–1	–1
p_z	Γ_3	+1	–1	–1	+1	–1	+1	+1	–1

Hence, it is found that each symmetry operation correspond to either +1 or –1 and these are the characters corresponding to the symmetry operations.

3.1.2 Matrix Representation of Unit Vectors as the Basis

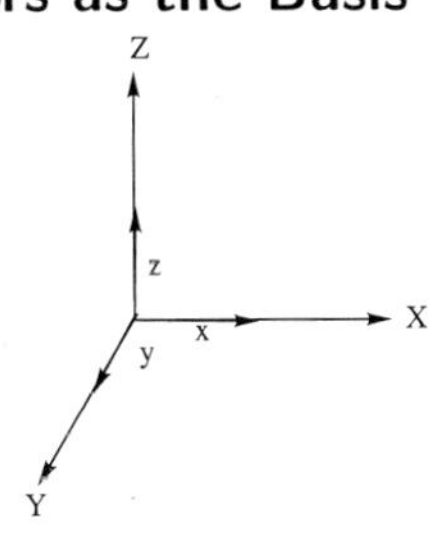

Let us take, the three unit vectors along the three Cartesian coordinates as basis and see the effect of different symmetry operations. Also let us assume that the unit vectors, x, y and z transform to x', y' and z' after symmetry operations, *i.e.*,

$E_x = x'$, $E_y = y'$ and $E_z = z'$

We can also write it in matrix form as,

$$E\begin{pmatrix} x \\ y \\ z \end{pmatrix} = \begin{pmatrix} x' \\ y' \\ z' \end{pmatrix} \quad i.e., \begin{matrix} x' & x & 0 & 0 \\ y' = 0 & y & 0 \\ z' & 0 & 0 & z \end{matrix} \text{ Or, } \mathbf{E} = \begin{pmatrix} 1 & 0 & 0 \\ 0 & 1 & 0 \\ 0 & 0 & 1 \end{pmatrix}$$

This 3×3 matrix is responsible to transform original x, y and z to new coordinate system, x', y' and z' corresponds to the symmetry operation, E.

Similarly, we can find the corresponding transformation matrix for $C_2(z)$ symmetry operation,

$$C_2(z)\begin{pmatrix}x\\y\\z\end{pmatrix}=\begin{pmatrix}x'\\y'\\z'\end{pmatrix}\ \textit{i.e.,}\ \begin{matrix}x' & -x & +0 & +0\\ y'= & 0 & -y & +0\\ z' & 0 & +0 & +z\end{matrix}\ \text{Or, } \mathbf{C_2(z)}=\begin{pmatrix}-1 & 0 & 0\\ 0 & -1 & 0\\ 0 & 0 & 1\end{pmatrix}$$

We can further try to find the matrix form of other symmetry operations as well:

$$\mathbf{C_2(x)}=\begin{pmatrix}1 & 0 & 0\\ 0 & -1 & 0\\ 0 & 0 & -1\end{pmatrix},\mathbf{C_2(y)}\begin{pmatrix}-1 & 0 & 0\\ 0 & 1 & 0\\ 0 & 0 & -1\end{pmatrix},\ \sigma_{\mathbf{xy}}=\begin{pmatrix}1 & 0 & 0\\ 0 & 1 & 0\\ 0 & 0 & -1\end{pmatrix}$$

$$\sigma_{\mathbf{yz}}=\begin{pmatrix}-1 & 0 & 0\\ 0 & 1 & 0\\ 0 & 0 & 1\end{pmatrix},\sigma_{\mathbf{xz}}=\begin{pmatrix}1 & 0 & 0\\ 0 & -1 & 0\\ 0 & 0 & 1\end{pmatrix},\ \mathbf{i}=\begin{pmatrix}-1 & 0 & 0\\ 0 & -1 & 0\\ 0 & 0 & 1\end{pmatrix}$$

3.1.3 Effect of C_n Axis (Rotational Angle, $\alpha = 2\pi/n$) on Unit Vectors

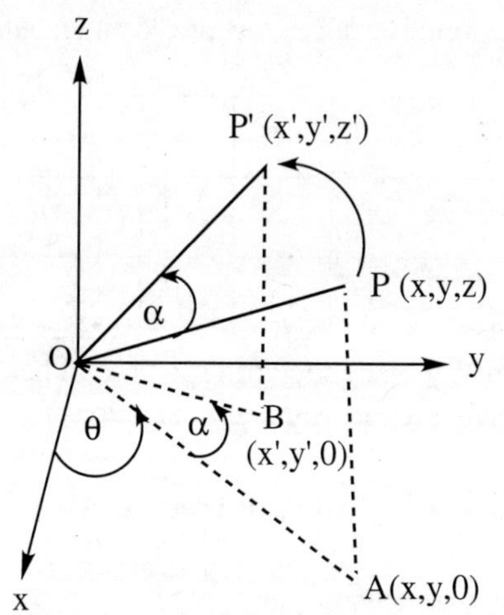

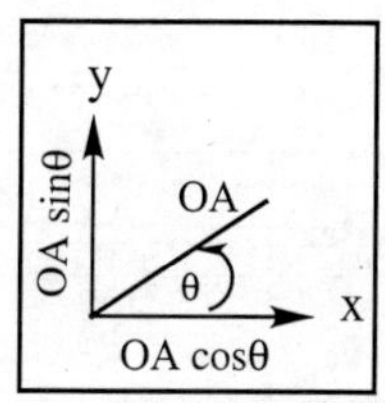

Let a point, $P(x,y,z)$ be rotated anti-clock wise about z-axis through an angle of α and the final position is $P'(x',y',z')$. $A(x,y,0)$ and $B(x',y',0)$ are the projections of P and P′ respectively on xy plane. Then,

$$x = \text{OA}\ \cos\theta \quad \text{and} \quad y = \text{OA}\ \sin\theta$$

Now,

$$x' = \text{OB}\ \cos(\theta+\alpha) = \text{OB}\ \cos\theta\cos\alpha - \text{OB}\ \sin\theta\sin\alpha$$
$$= x\ \cos\alpha - y\ \sin\alpha\ (\text{as OA} = \text{OB})$$

Thus,

$$x' = x\ \cos\alpha - y\ \sin\alpha + z.0 \tag{3.1}$$

Similarly,

$$y' = \text{OB } \sin(\theta+\alpha) = \text{OB } \sin\theta\cos\alpha - \text{OB } \cos\theta\sin\alpha$$
$$= y \cos\alpha + x \sin\alpha \qquad (\text{as OA} = \text{OB})$$

Or,

$$y' = y \cos\alpha + x \sin\alpha + z.0 \tag{3.2}$$

And also

$$z' = z, \textit{ i.e.}, z' = x.0 + y.0 + z.0 \tag{3.3}$$

From equations (3.1)–(3.3), we can write in matrix form as

$$\begin{pmatrix} x' & y' & z' \end{pmatrix} = \begin{pmatrix} x & y & z \end{pmatrix} \begin{pmatrix} \cos\alpha & \sin\alpha & 0 \\ -\sin\alpha & \cos\alpha & 0 \\ 0 & 0 & 1 \end{pmatrix}$$

Hence, the matrix representation of $C_\alpha(z)$ (rotational operator about z-axis through an angle of α operation with the three unit vectors as basis is

$$C_n = \begin{pmatrix} \cos\alpha & \sin\alpha & 0 \\ -\sin\alpha & \cos\alpha & 0 \\ 0 & 0 & 1 \end{pmatrix}; \text{ where } \quad \alpha = 2\pi/n$$

3.1.4 Matrix Representation of Reflection Operation with Three Unit Vectors as Basis

Let MN be the reflection plane on xy plane and a point, P at space transform to P′ after reflection. We have $x = r\cos\theta$ and $y = r\sin\theta$. Now,

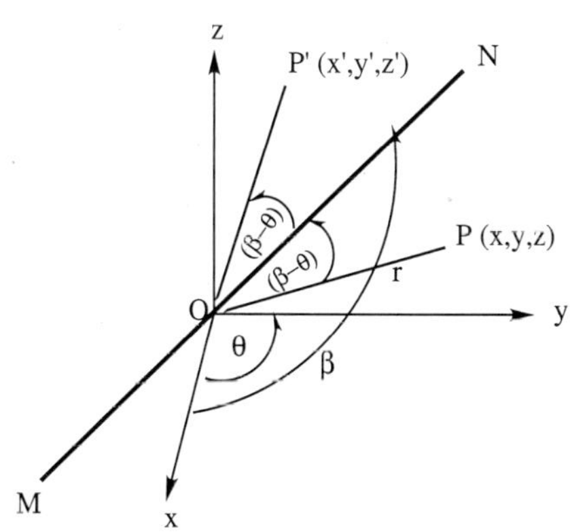

$$x' = r \cos[\theta + (\beta-\theta) + (\beta-\theta)]$$
$$= r \cos(2\beta - \theta)$$

$$= r\cos 2\beta\cos\theta + r\sin 2\beta\sin\theta$$
$$= x\cos 2\beta + y\sin 2\beta + z.0$$

Again,

$$y' = r\sin[\theta+(\beta-\theta)+(\beta-\theta)] = r\sin(2\beta-\theta)$$
$$= x\sin 2\beta - y\cos 2\beta + z.0,$$

and also, $z' = x.0 + y.0 + z.1$
Hence,

$$\begin{pmatrix} x' & y' & z' \end{pmatrix} = \begin{pmatrix} x & y & z \end{pmatrix} \begin{pmatrix} \cos 2\beta & \sin 2\beta & 0 \\ \sin 2\beta & -\cos 2\beta & 0 \\ 0 & 0 & 1 \end{pmatrix}$$

Hence, the matrix representation of σ operation with the three unit vectors as basis is

$$\sigma = \begin{pmatrix} \cos 2\beta & \sin 2\beta & 0 \\ \sin 2\beta & -\cos 2\beta & 0 \\ 0 & 0 & 1 \end{pmatrix}$$

3.1.5 Matrix Representation of Improper Rotation (S_n) with Three Unit Vectors as Basis

Here C_n axis is along the z-axis and the σ_{xy} is perpendicular to the C_n axis. We know that $S_n = \sigma_{xy}C_n$

$$S_n \begin{pmatrix} x \\ y \\ z \end{pmatrix} = \begin{pmatrix} x' \\ y' \\ z' \end{pmatrix} \text{ i.e., } \sigma_{xy}C_n \begin{pmatrix} x \\ y \\ z \end{pmatrix} = \begin{pmatrix} x' \\ y' \\ z' \end{pmatrix} \text{ i.e.,}$$

We already have,

$$C_n = \begin{pmatrix} \cos\alpha & \sin\alpha & 0 \\ -\sin\alpha & \cos\alpha & 0 \\ 0 & 0 & 1 \end{pmatrix} \text{ and } \sigma_{xy} = \begin{pmatrix} 1 & 0 & 0 \\ 0 & 1 & 0 \\ 0 & 0 & -1 \end{pmatrix};$$

Hence,

$$S_n = \begin{pmatrix} \cos\alpha & \sin\alpha & 0 \\ -\sin\alpha & \cos\alpha & 0 \\ 0 & 0 & 1 \end{pmatrix} \begin{pmatrix} 1 & 0 & 0 \\ 0 & 1 & 0 \\ 0 & 0 & -1 \end{pmatrix} = \begin{pmatrix} \cos\alpha & \sin\alpha & 0 \\ -\sin\alpha & \cos\alpha & 0 \\ 0 & 0 & -1 \end{pmatrix}$$

3.1.6 Matrix Representation of Inversion Operation (i) with Three Unit Vectors as Basis

$$x' = -x, \text{ i.e., } x' = -1.x + 0.y + 0.z$$
$$y' = -y, \text{ i.e., } x' = 0.x + (-1).y + 0.z$$
$$z' = -z, \text{ i.e., } x' = 0.x + 0.y + (-1).z$$

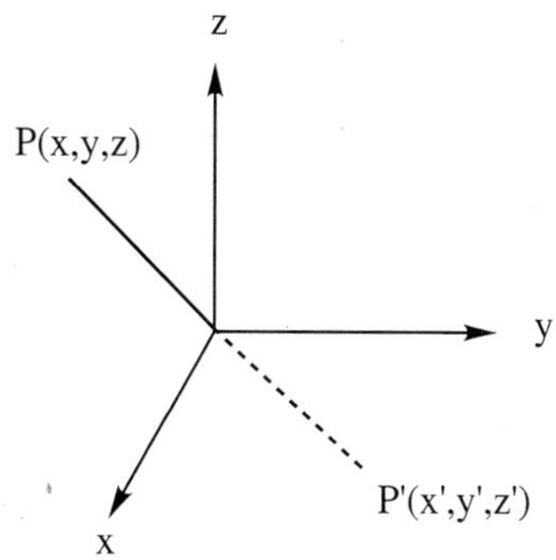

Hence, the corresponding matrix representation of inversion operation with three unit vectors as basis will be

$$i = \begin{pmatrix} -1 & 0 & 0 \\ 0 & -1 & 0 \\ 0 & 0 & -1 \end{pmatrix}$$

3.1.7 Rotational Vectors as Basis for Representation

From the figure, it is evident that R_x vector is parallel to σ_{yz} plane, R_y is parallel to σ_{xz} plane and R_z is parallel to σ_{xy} plane; When an arrow after striking a plane reversed its arrow head, it is antisymmetric with respect to that operation. For example, R_z is antisymmetric with respect to σ_{xz} and σ_{yz}. Again R_x and R_y are antisymmetric with respect to the $C_2(z)$ operation.

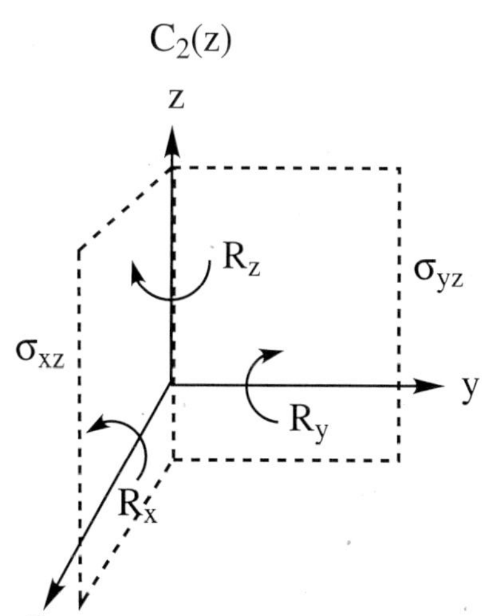

The characters can be represented in a tabular form for the symmetry operations corresponding to C_{2v} point group as follows:

C_{2v}	E	$C_2(z)$	σ_{xz}	σ_{yz}
R_x	1	–1	–1	1
R_y	1	–1	1	–1
R_z	1	1	–1	–1

We have learnt how to get representation for any symmetry operation using some basis. Now, remember the first example, where the P_x, P_y and P_z orbitals were taken as basis and the symmetry operations namely E, $C_2(z)$, σ_{xz} and σ_{yz} , corresponding to C_{2v} point group is considered. Along with those representations, here we will add another row taking R_z as basis and the full list of characters can be tabulated as follows:

C_{2v}	E	$C_2(z)$	σ_{yz}	σ_{xz}	
A_1	1	1	1	1	P_z
A_2	1	1	–1	–1	R_z
B_1	1	–1	1	–1	P_x
B_2	1	–1	–1	1	P_y

Here, A_1, A_2, B_1 and B_2 are the names of a particular representation. At this point do not worry about the nomenclature of those representation, as it has been discussed later. We say that x or P_x orbital is symmetric with respect to E and σ_{yz} operations and antisymmetric with respect to C_2 and σ_{xz} operations. Similar kind of conclusions can drawn for other basis present here. But here A_1 is called **totally symmetric** representation as all the characters are +1 irrespective of nature of symmetry operation. Hence, it is said that P_z or z axis belongs to the totally symmetric representation of C_2v point group. In this way the full set of representation along with their basis (at the last column) included in a table is called the Character table of C_{2v} point group. Later it has been discussed in detail in a separate chapter. Here each of the representation is called the **Irreducible representation** (IR) as these representations can not be reduced further.

3.2 Reducible Representation

So far we have seen that characters of irreducible representations (IR) can not be further reduced (as denoted by either or 1 or –1). There is however, another kind of representation which can be reduced further to its corresponding component IRs, called **Reducible representation**.

In group theory, to solve any real problem it involves generation a reducible representation, followed by reduction of the reducible representation to its constituent IRs. It is not possible to get the IRs directly for a real problem.

For example, let 3 –1 –1 –1 is a reducible representation of C_{2v} point group.

After a few trial and error methods, by various combinations of the IRs present in the character table of C_{2v}, it can be found as:

C_{2v}	E	$C_{2v}(z)$	σ_{yz}	σ_{xz}
Γ_1	3	–1	–1	–1

A_2	1	1	–1	–1
B_1	1	–1	1	–1
B_2	1	–1	–1	1
So, $A_2 + B_1 + B_2$	3	–1	–1	–1

Another example can be taken. Let, 3 3 1 1 is a reducible representation for C_{2v} point group and combination of various IRs can be constructed in a similar manner.

A_1	1	1	1	1
A_1	1	1	1	1
A_2	1	1	–1	1
Thus, $2A_1 + A_2$	3	1	1	1

But this is a trial and error method. We have a Reduction formula and using that formula we can very easily find the component irreducible representations of any reducible representation.

3.2.1 Generation of Reducible Representation

Let us take the example of H_2O molecule and each atom is taken as origin, positioning three vectors along the cartesian coordinates x,y,z.

The molecule transform after the four operations under C_{2v} point group according to the scheme presented below and let us construct the matrix representation for each operation. We designate the vectors by prime (x′, y′ and z′) after transformation against each symmetry operation.

The matrix equation corresponding to identity operation (E) can be written as:

$$E\begin{pmatrix}x_1\\y_1\\z_1\\x_2\\y_2\\z_2\\x_3\\y_3\\z_3\end{pmatrix}=\begin{pmatrix}x'_1\\y'_1\\z'_1\\x'_2\\y'_2\\z'_2\\x'_3\\y'_3\\z'_3\end{pmatrix}\quad i.e.,\begin{array}{lllllllll}
x'_1=x_1 & +0 & +0 & +0 & +0 & +0 & +0 & +0 & +0\\
y'_1=0 & +y_1 & +0 & +0 & +0 & +0 & +0 & +0 & +0\\
z'_1=0 & +0 & +z_1 & +0 & +0 & +0 & +0 & +0 & +0\\
x'_2=0 & +0 & +0 & +x_2 & +0 & +0 & +0 & +0 & +0\\
y'_2=0 & +0 & +0 & +0 & +y_2 & +0 & +0 & +0 & +0\\
z'_2=0 & +0 & +0 & +0 & +0 & +z_2 & +0 & +0 & +0\\
x'_3=0 & +0 & +0 & +0 & +0 & +0 & +x_3 & +0 & +0\\
y'_3=0 & +0 & +0 & +0 & +0 & +0 & +0 & +y_3 & +0\\
z'_3=0 & +0 & +0 & +0 & +0 & +0 & +0 & +0 & +z_3
\end{array}$$

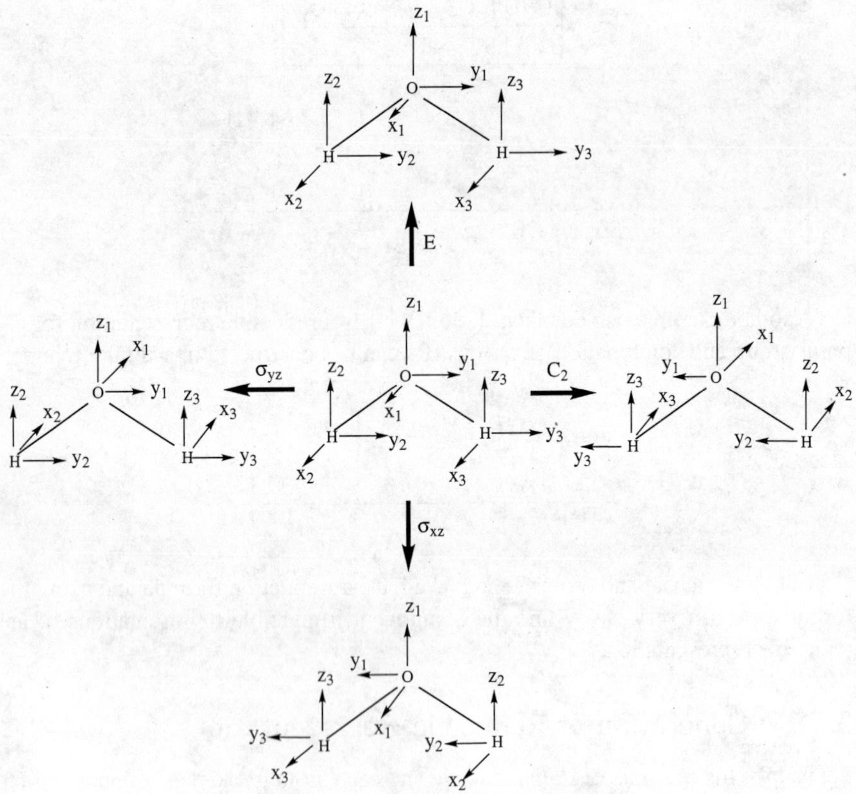

Hence, the matrix corresponding to operation E becomes as:

$$
\begin{pmatrix}
1 & 0 & 0 & 0 & 0 & 0 & 0 & 0 & 0 \\
0 & 1 & 0 & 0 & 0 & 0 & 0 & 0 & 0 \\
0 & 0 & 1 & 0 & 0 & 0 & 0 & 0 & 0 \\
0 & 0 & 0 & 1 & 0 & 0 & 0 & 0 & 0 \\
0 & 0 & 0 & 0 & 1 & 0 & 0 & 0 & 0 \\
0 & 0 & 0 & 0 & 0 & 1 & 0 & 0 & 0 \\
0 & 0 & 0 & 0 & 0 & 0 & 1 & 0 & 0 \\
0 & 0 & 0 & 0 & 0 & 0 & 0 & 1 & 0 \\
0 & 0 & 0 & 0 & 0 & 0 & 0 & 0 & 1
\end{pmatrix}
\begin{pmatrix}
x_1 \\ y_1 \\ z_1 \\ x_2 \\ y_2 \\ z_2 \\ x_3 \\ y_3 \\ z_3
\end{pmatrix}
$$

When we add the diagonal elements of the matrix we get trace of the matrix or character corresponding to that symmetry operation. Here, Trace, $\chi = 9$

Similarly, **for** $\mathbf{C}_2$ symmetry operation, the matrix equation can be written as:

$$\mathbf{C}_2 \begin{pmatrix} x_1 \\ y_1 \\ z_1 \\ x_2 \\ y_2 \\ z_2 \\ x_3 \\ y_3 \\ z_3 \end{pmatrix} = \begin{pmatrix} x_1' \\ y_1' \\ z_1' \\ x_2' \\ y_2' \\ z_2' \\ x_3' \\ y_3' \\ z_3' \end{pmatrix} \quad i.e., \begin{array}{rcccccccc} x_1' = -x_1 & +0 & +0 & +0 & +0 & +0 & +0 & +0 & +0 \\ y_1' = 0 & -y_1 & +0 & +0 & +0 & +0 & +0 & +0 & +0 \\ z_1' = 0 & +0 & +z_1 & +0 & +0 & +0 & +0 & +0 & +0 \\ x_2' = 0 & +0 & +0 & +0 & +0 & +0 & -x_3 & +0 & +0 \\ y_2' = 0 & +0 & +0 & +0 & +0 & +0 & +0 & -y_3 & +0 \\ z_2' = 0 & +0 & +0 & +0 & +0 & +0 & +0 & +0 & +z_3 \\ x_3' = 0 & +0 & +0 & -x_2 & +0 & +0 & +0 & +0 & +0 \\ y_3' = 0 & +0 & +0 & +0 & -y_2 & +0 & +0 & +0 & +0 \\ z_3' = 0 & +0 & +0 & +0 & +0 & +z_2 & +0 & +0 & +0 \end{array}$$

Hence, the matrix corresponding to operation C_2 will be:

$$\begin{pmatrix} -1 & 0 & 0 & 0 & 0 & 0 & 0 & 0 & 0 \\ 0 & -1 & 0 & 0 & 0 & 0 & 0 & 0 & 0 \\ 0 & 0 & 1 & 0 & 0 & 0 & 0 & 0 & 0 \\ 0 & 0 & 0 & 0 & 0 & 0 & -1 & 0 & 0 \\ 0 & 0 & 0 & 0 & 0 & 0 & 0 & -1 & 0 \\ 0 & 0 & 0 & 0 & 0 & 0 & 0 & 0 & 1 \\ 0 & 0 & 0 & -1 & 0 & 0 & 0 & 0 & 0 \\ 0 & 0 & 0 & 0 & -1 & 0 & 0 & 0 & 0 \\ 0 & 0 & 0 & 0 & 0 & 1 & 0 & 0 & 0 \end{pmatrix} \begin{pmatrix} x_1 \\ y_1 \\ z_1 \\ x_2 \\ y_2 \\ z_2 \\ x_3 \\ y_3 \\ z_3 \end{pmatrix}$$

$$\text{Trace}(\chi) = -1$$

Matrix equation **for** $\sigma_{\mathbf{xz}}$-operation:

$$\sigma_{xz} \begin{pmatrix} x_1 \\ y_1 \\ z_1 \\ x_2 \\ y_2 \\ z_2 \\ x_3 \\ y_3 \\ z_3 \end{pmatrix} = \begin{pmatrix} x_1' \\ y_1' \\ z_1' \\ x_2' \\ y_2' \\ z_2' \\ x_3' \\ y_3' \\ z_3' \end{pmatrix} \quad i.e., \begin{array}{rcccccccc} x_1' = x_1 & +0 & +0 & +0 & +0 & +0 & +0 & +0 & +0 \\ y_1' = 0 & -y_1 & +0 & +0 & +0 & +0 & +0 & +0 & +0 \\ z_1' = 0 & +0 & +z_1 & +0 & +0 & +0 & +0 & +0 & +0 \\ x_2' = 0 & +0 & +0 & +0 & +0 & +0 & +x_3 & +0 & +0 \\ y_2' = 0 & +0 & +0 & +0 & +y_2 & +0 & +0 & -y_3 & +0 \\ z_2' = 0 & +0 & +0 & +0 & +0 & +z_2 & +0 & +0 & +z_3 \\ x_3' = 0 & +0 & +0 & +x_2 & +0 & +0 & +0 & +0 & +0 \\ y_3' = 0 & +0 & +0 & +0 & -y_2 & +0 & +0 & +0 & +0 \\ z_3' = 0 & +0 & +0 & +0 & +0 & +z_2 & +0 & +0 & +0 \end{array}$$

Hence, the matrix corresponding to operation σ_{xz} will be:

$$\begin{pmatrix} 1 & 0 & 0 & 0 & 0 & 0 & 0 & 0 & 0 \\ 0 & -1 & 0 & 0 & 0 & 0 & 0 & 0 & 0 \\ 0 & 0 & 1 & 0 & 0 & 0 & 0 & 0 & 0 \\ 0 & 0 & 0 & 0 & 0 & 0 & 1 & 0 & 0 \\ 0 & 0 & 0 & 0 & 0 & 0 & 0 & -1 & 0 \\ 0 & 0 & 0 & 0 & 0 & 0 & 0 & 0 & 1 \\ 0 & 0 & 0 & 1 & 0 & 0 & 0 & 0 & 0 \\ 0 & 0 & 0 & 0 & -1 & 0 & 0 & 0 & 0 \\ 0 & 0 & 0 & 0 & 0 & 1 & 0 & 0 & 0 \end{pmatrix} \begin{pmatrix} x_1 \\ y_1 \\ z_1 \\ x_2 \\ y_2 \\ z_2 \\ x_3 \\ y_3 \\ z_3 \end{pmatrix}$$

$$\text{Trace}(\chi) = 1$$

Matrix equation **for σ_{yz}**-operation:

$$\sigma_{yz} \begin{pmatrix} x_1 \\ y_1 \\ z_1 \\ x_2 \\ y_2 \\ z_2 \\ x_3 \\ y_3 \\ z_3 \end{pmatrix} = \begin{pmatrix} x_1' \\ y_1' \\ z_1' \\ x_2' \\ y_2' \\ z_2' \\ x_3' \\ y_3' \\ z_3' \end{pmatrix} \quad i.e., \quad \begin{array}{lllllllll} x_1' = -x_1 & +0 & +0 & +0 & +0 & +0 & +0 & +0 & +0 \\ y_1' = 0 & +y_1 & +0 & +0 & +0 & +0 & +0 & +0 & +0 \\ z_1' = 0 & +0 & +z_1 & +0 & +0 & +0 & +0 & +0 & +0 \\ x_2' = 0 & +0 & +0 & -x_2 & +0 & +0 & +0 & +0 & +0 \\ y_2' = 0 & +0 & +0 & +0 & +y_2 & +0 & +0 & +0 & +0 \\ z_2' = 0 & +0 & +0 & +0 & +0 & +z_2 & +0 & +0 & +0 \\ x_3' = 0 & +0 & +0 & +0 & +0 & +0 & -x_3 & +0 & +0 \\ y_3' = 0 & +0 & +0 & +0 & +0 & +0 & +0 & +y_3 & +0 \\ z_3' = 0 & +0 & +0 & +0 & +0 & +0 & +0 & +0 & +z_3 \end{array}$$

Thus, the matrix corresponding to operation σ_{yz} will be

$$\begin{pmatrix} -1 & 0 & 0 & 0 & 0 & 0 & 0 & 0 & 0 \\ 0 & 1 & 0 & 0 & 0 & 0 & 0 & 0 & 0 \\ 0 & 0 & 1 & 0 & 0 & 0 & 0 & 0 & 0 \\ 0 & 0 & 0 & -1 & 0 & 0 & 0 & 0 & 0 \\ 0 & 0 & 0 & 0 & 1 & 0 & 0 & 0 & 0 \\ 0 & 0 & 0 & 0 & 0 & 1 & 0 & 0 & 0 \\ 0 & 0 & 0 & 0 & 0 & 0 & -1 & 0 & 0 \\ 0 & 0 & 0 & 0 & 0 & 0 & 0 & 1 & 0 \\ 0 & 0 & 0 & 0 & 0 & 0 & 0 & 0 & 1 \end{pmatrix} \begin{pmatrix} x_1 \\ y_1 \\ z_1 \\ x_2 \\ y_2 \\ z_2 \\ x_3 \\ y_3 \\ z_3 \end{pmatrix}$$

$$\text{Trace}(\chi) = 3$$

Thus, we have obtained 9×9 matrices corresponding to each symmetry operation of C_{2v} point group using total 9 unit vectors as basis (three on each atom). Taking cartesian coordinate as basis, if we take the trace of each matrix corresponding to each operation, then we get the reducible representation of that point group as shown below in the table:

C_{2v}	E	C_2	σ_{xz}	σ_{yz}
Γ_{3N}	9	−1	1	3

This reducible representation, Γ_{3N} can further be reduced into component irreducible representations. As already stated that irreducible representations can not be further reduced and these representations are very important to analyze any problem in terms of group theory. This reduction procedure is very cumbersome and time consuming and for larger molecules this becomes more complicated. For example, if we want to find reducible representation for a molecule with 12 atoms like, C_6H_6 molecule, then we have to deal with a 36×36 matrix. So we need simplified procedure.

From 3N unit vectors it is clear that the vectors or atoms which are shifted during symmetry operation, they appear in off-diagonal position of matrix and hence, do not contribute to the overall character or trace of the matrix. Only unshifted atoms or vectors have contribution since they always appear in the diagonal position of a matrix. Thus, we can find out the reducible representation corresponding to each symmetry operation from only the consideration of number of unshifted atoms using some basis.

Hence, without generating the matrices, simply we can consider the number of unshifted basis with each operation and we can get the representation directly. Here the basis is the three vectors along the three axis taking each atom as origin as shown in previous figure for water molecule. We can tabulate the information and find the character for each operation as follows:

C_{2v}	**E**	C_2	σ_{xz}	σ_{yz}
No. of unshifted unit vectors after symmetry operation	**9** (All the 9 unit vectors on 3 atoms remain unshifted)	**−1** (Except z_1 of O-atom, all other vectors change their position. z_1 also changes its sign, so −1)	**1** (z_1 and x_1 of O-atom remain unchanged and y_1 changes its sign. All other vectors change their position, so 2−1=1)	**3** (All y and z vectors remain unchanged and x's change their sign. So 6−3=3)

3.2.1.1 Reducible Representations Generated Using Three Unit Vectors as Basis

1. **With identity operation, E**

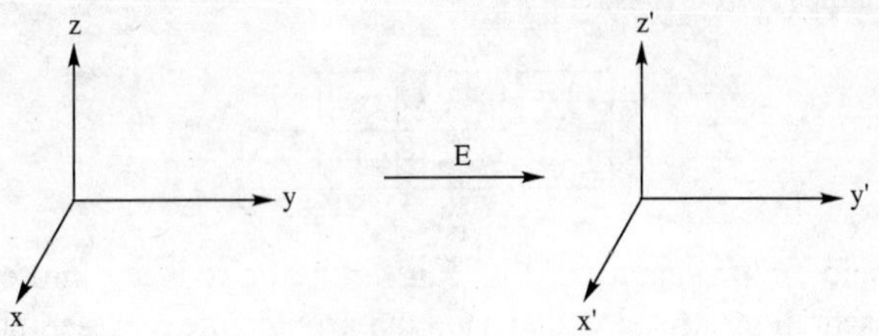

$$E\begin{pmatrix}x\\y\\z\end{pmatrix}=\begin{pmatrix}x'\\y'\\z'\end{pmatrix}\quad i.e.,\quad \begin{matrix}x'=x & +0 & +0\\ y'=0 & +y & +0\\ z'=0 & +0 & +z\end{matrix}\quad \text{Or,}\quad E=\begin{pmatrix}1&0&0\\0&1&0\\0&0&1\end{pmatrix}$$

$$i.e.\quad \chi=3$$

2. **Inversion operation, *i***

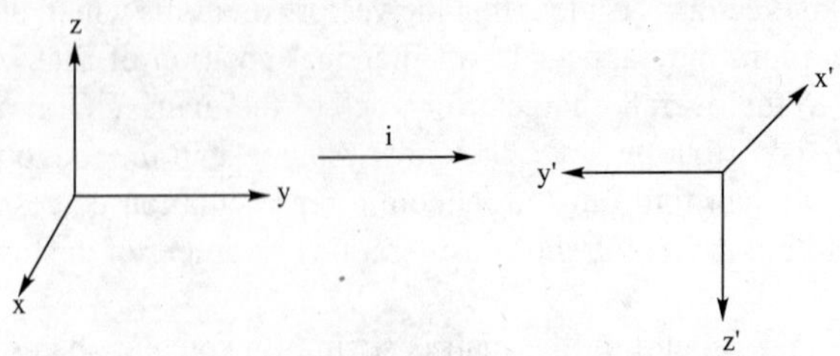

$$i\begin{pmatrix}x\\y\\z\end{pmatrix}=\begin{pmatrix}x'\\y'\\z'\end{pmatrix}\quad i.e.,\quad \begin{matrix}x'=-x & +0 & +0\\ y'=0 & -y & +0\\ z'=0 & +0 & -z\end{matrix}\quad \text{Or,}\quad i=\begin{pmatrix}-1&0&0\\0&-1&0\\0&0&-1\end{pmatrix}$$

$$i.e.\quad \chi=-3$$

3. **Reflection operation, σ**

$$\sigma_{xz}\begin{pmatrix}x\\y\\z\end{pmatrix}=\begin{pmatrix}x'\\y'\\z'\end{pmatrix}\quad i.e.,\quad \begin{matrix}x'=x & +0 & +0\\ y'=0 & -y & +0\\ z'=0 & +0 & +z\end{matrix}\quad \text{Or,}\quad \sigma_{xz}=\begin{pmatrix}1&0&0\\0&-1&0\\0&0&1\end{pmatrix}$$

$$i.e.\quad \chi=1$$

$$\sigma_{yz}\begin{pmatrix}x\\y\\z\end{pmatrix}=\begin{pmatrix}x'\\y'\\z'\end{pmatrix}\quad i.e.,\quad \begin{matrix}x'=-x & +0 & +0\\ y'=0 & y & +0\\ z'=0 & +0 & z\end{matrix}\quad \text{Or,}\quad \sigma_{yz}=\begin{pmatrix}-1&0&0\\0&1&0\\0&0&1\end{pmatrix}$$

$$i.e.\quad \chi=1$$

$$\sigma_{h}\begin{pmatrix}x\\y\\z\end{pmatrix}=\begin{pmatrix}x'\\y'\\z'\end{pmatrix}\quad i.e.,\quad \begin{matrix}x'=x & +0 & +0\\ y'=0 & +y & +0\\ z'=0 & +0 & -z\end{matrix}\quad \text{Or,}\quad \sigma_{h}=\begin{pmatrix}1&0&0\\0&1&0\\0&0&-1\end{pmatrix}$$

$$i.e.\quad \chi=1$$

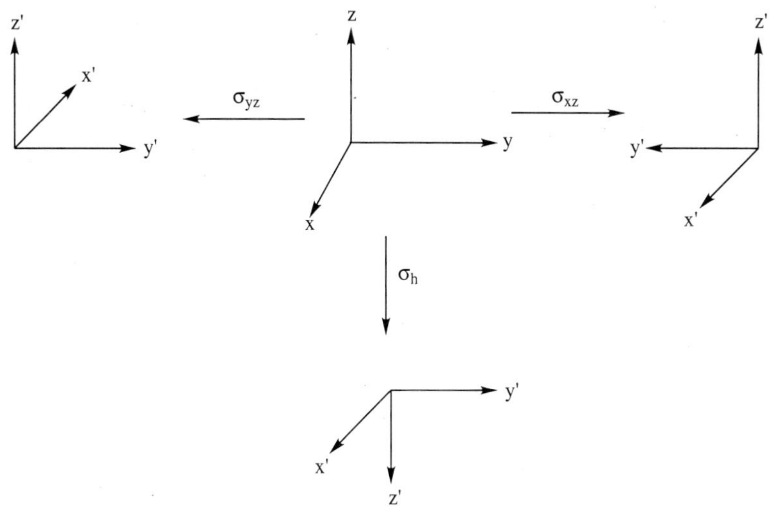

4. **C_2 operation along z-axis**

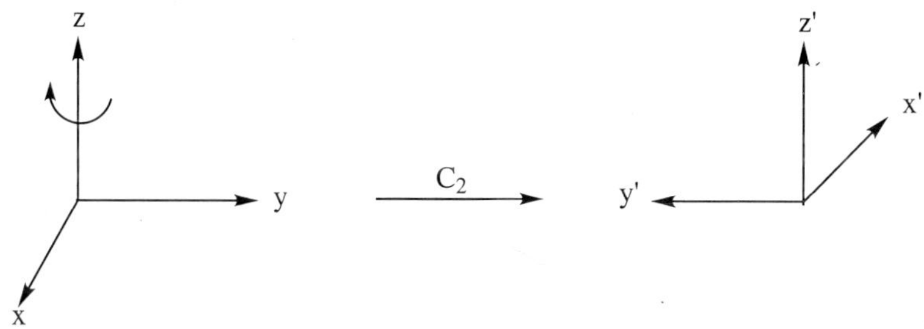

$$C_2\begin{pmatrix}x\\y\\z\end{pmatrix}=\begin{pmatrix}x'\\y'\\z'\end{pmatrix}\quad i.e.,\quad \begin{matrix}x'=-x & +0 & +0\\ y'=0 & -y & +0\\ z'=0 & +0 & z\end{matrix}\quad \text{Or,}\quad C_2=\begin{pmatrix}-1&0&0\\0&-1&0\\0&0&1\end{pmatrix}$$

$$i.e.\quad \chi=-1$$

Alternatively, by counting the number of shifted and unshifted atoms we can make the table.

No. of unshifted/unshifted atoms	E	i	C_2	σ_{xy}	σ_{yz}	σ_{yz}
Γ_{3N}	3	–3	–1	1	1	1

We have the general table of character per unshifted atom using 3N unit vectors as the basis as shown below:

Operation, R	Character, χ
E	+3
i	–3
σ	+1
C_2	–1
C_3^1, C_3^2	0
C_4^1, C_4^3	+1
C_6^1, C_6^5	+2
S_3^1, S_3^5	–2
S_4^1, S_4^3	–1
S_6^1, S_6^5	0
Proper axis (C_n)	$2\cos\theta + 1$
Improper axis (S_n)	$2\cos\theta - 1$

3.2.2 Example

3.2.2.1 Reducible Representation for C_{2v} Point Group

The number of unshifted atoms for water molecule against each operation under C_{2v} point group can easily be found from the picture below.

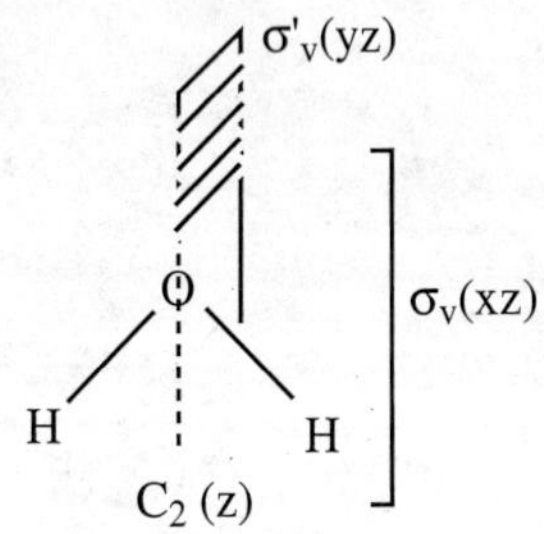

1. E does not shift any atom, hence, the number of unshifted atoms are 3.
2. C_2 does not shift O atom but interchanges two H atoms, hence unshifted atom is 1.
3. σ_{xz} or σ_v does not shift O atom but interchanges two H atoms, hence unshifted atom is 1.
4. σ_{yz} or σ_v does not shift any atom, hence number of unshifted atoms are 3.

Therefore, we can write the reducible representation for the water molecule taking the three unit vector as basis as follows in the table below:

It is interesting to find that these characters are same as we have found earlier from the summation of diagonal elements of the matrices corresponding to each symmetry operation.

C_{2v}	E	C_2	σ_{xz}	σ_{yz}
No. of unshifted atoms by symmetry operations	9	1	1	3
Γ_{3N} (No. unshifted atom × Character of R)	$3 \times 3 = 9$	$1 \times -1 = -1$	$1 \times 1 = 1$	$3 \times 1 = 3$

3.2.2.2 Reducible Representations for Tetrahedral Geometry Using Bond Vectors as Basis

Methane molecule has been taken as example for discussion. Here the molecule has been represented in two different orientations as (I) and (II).

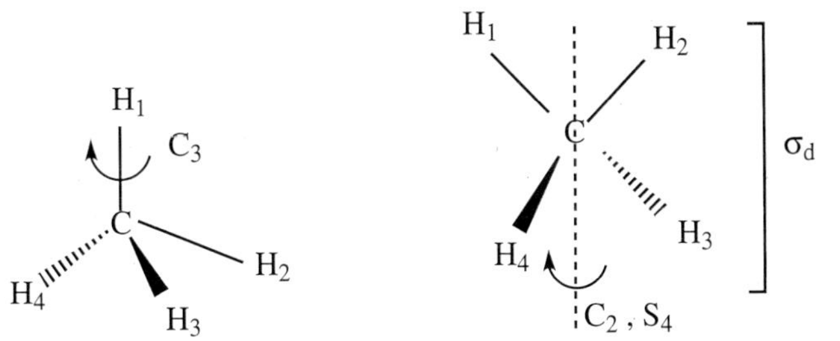

(I) (II)

Figure 3.2: Different symmetry operations on tetrahedral geometry

σ_d is in plane of the paper and S_4 is collinear with C_2.

There are four C-H bond vectors (B_1, B_2, B_3 and B_4) in CH_4. After the operation let the corresponding new bond vectors becomes B'_1, B'_2, B'_3 and B'_4. Now we will try both the methods to construct reducible representation using those four bond vectors as basis.

Method 1:

It is found that on application of E operation on various bond vectors, CH_1 bond (B_1) remains CH_1 bond, CH_2 bond (B_2) remains CH_2 bond, CH_3 bond (B_3) remains CH_3 bond, CH_4 bond (B_4) remains CH_4 bond, *i.e.*,

$$E \begin{pmatrix} B_1 \\ B_2 \\ B_3 \\ B_4 \end{pmatrix} = \begin{pmatrix} B'_1 \\ B'_2 \\ B'_3 \\ B'_4 \end{pmatrix} \quad i.e., \quad \begin{matrix} B'_1 = B_1 & +0 & +0 & +0 \\ B'_2 - 0 & +B_2 & +0 & |\,0 \\ B'_3 = 0 & +0 & +B_3 & +0 \\ B'_4 = 0 & +0 & +0 & +B_4 \end{matrix} \quad \text{Or,} \quad E = \begin{pmatrix} 1 & 0 & 0 & 0 \\ 0 & 1 & 0 & 0 \\ 0 & 0 & 1 & 0 \\ 0 & 0 & 0 & 1 \end{pmatrix}$$

Charater= 4.

For C_3,

B_1 remains B_1, B_2 changes to B'_3, B_3 changes to B'_4 and B_4 changes to B'_2.

$$C_3 \begin{pmatrix} B_1 \\ B_2 \\ B_3 \\ B_4 \end{pmatrix} = \begin{pmatrix} B'_1 \\ B'_2 \\ B'_3 \\ B'_4 \end{pmatrix} \quad i.e., \quad \begin{matrix} B'_1 = B_1 & +0 & +0 & +0 \\ B'_2 = 0 & +0 & +B_3 & +0 \\ B'_3 = 0 & +0 & +0 & +B_4 \\ B'_4 = 0 & +B_2 & +0 & +0 \end{matrix}$$

Hence, the matrix can be written as

$$\begin{pmatrix} B'_1 \\ B'_2 \\ B'_3 \\ B'_4 \end{pmatrix} = \begin{pmatrix} B_1 \\ B_2 \\ B_3 \\ B_4 \end{pmatrix} \begin{pmatrix} 1 & 0 & 0 & 0 \\ 0 & 0 & 1 & 0 \\ 0 & 0 & 0 & 1 \\ 0 & 1 & 0 & 0 \end{pmatrix}$$

Hence, the matrix representation corresponding to C_3 operation is as follow with character =1

$$C_3 = \begin{pmatrix} 1 & 0 & 0 & 0 \\ 0 & 0 & 1 & 0 \\ 0 & 0 & 0 & 1 \\ 0 & 1 & 0 & 0 \end{pmatrix}$$

For C_2, B_1 becomes B'_2, B_2 changes to B'_1, B_3 changes to B'_4 and B_4 changes to B'_3.

$$C_2 \begin{pmatrix} B_1 \\ B_2 \\ B_3 \\ B_4 \end{pmatrix} = \begin{pmatrix} B'_1 \\ B'_2 \\ B'_3 \\ B'_4 \end{pmatrix} \quad i.e., \quad \begin{matrix} B'_1 = 0 & +B_2 & +0 & +0 \\ B'_2 = B_1 & +0 & +0 & +0 \\ B'_3 = 0 & +0 & +0 & +B_4 \\ B'_4 = 0 & +0 & +B_3 & +0 \end{matrix} \quad \text{Or, } C_2 = \begin{pmatrix} 0 & 1 & 0 & 0 \\ 1 & 0 & 0 & 0 \\ 0 & 0 & 0 & 1 \\ 0 & 0 & 1 & 0 \end{pmatrix}$$

The character of the matrix is 0.

For σ_d, B_1 remains B'_1, B_2 remains to B'_2, B_3 changes to B'_4 and B_4 changes to B'_3, *i.e.,*

$$\sigma_d \begin{pmatrix} B_1 \\ B_2 \\ B_3 \\ B_4 \end{pmatrix} = \begin{pmatrix} B'_1 \\ B'_2 \\ B'_3 \\ B'_4 \end{pmatrix} \quad i.e., \quad \begin{matrix} B'_1 = B_1 & +0 & +0 & +0 \\ B'_2 = 0 & +B_2 & +0 & +0 \\ B'_3 = 0 & +0 & +0 & +B_3 \\ B'_4 = 0 & +0 & +B_4 & +0 \end{matrix} \quad \text{Or, } \sigma_d = \begin{pmatrix} 1 & 0 & 0 & 0 \\ 0 & 1 & 0 & 0 \\ 0 & 0 & 0 & 1 \\ 0 & 0 & 1 & 0 \end{pmatrix}$$

For this matrix, the character is 2.

Similarly, for S_4 symmetry element, the characters can be found out. The characters corresponding to all the symmetry operations can be tabulated as:

T_d	E	$8C_3$	$3C_2$	$6S_4$	$6\sigma_d$
Γ_1	4	1	0	0	2

Method 2:

Let us follow the simple rules mentioned earlier and find the characters of all the symmetry elements for T_d molecule:

1. The bonds that are shifted by symmetry operations will contribute zero to the character of reducible representation.
2. The bonds that are not shifted contribute +1 to the character of reducible representation.
3. The bond vectors that are not shifted, but its direction is reversed will contribute –1.

T_d	E	$8C_3$	$3C_2$	$6S_4$	$6\sigma_d$
No. of unshifted bond vectors by symmetry operation	**4** (All the 4 bond vectors remain unshifted	**1** (Only B_1 remains unshifted)	**0** (All bonds changed)	**0** (All bonds changed)	**2** (B_1 and B_2 remains unchanged and other two bonds changed their position)

3.2.2.3 Reducible Representations Generated Using *p*-Orbitals as Basis for *cis*-1, 3-Butadiene

Let us consider the example of cis-1,3-butadiene which is planar molecule (C_{2v} point group). In this molecule, each C-atom has an unhybridized *p*-orbital and the four unhybridized *p*-orbitals are taken as the bases for the representation. Let us consider that ϕ_1, ϕ_2, ϕ_3 and ϕ_4 are the original configurations of unhybridized p orbitals and ϕ_1', ϕ_2', ϕ_3' and ϕ_4' be the new configuration of the orbitals after performing the symmetry operation corresponding to the point group. Although the molecule is planar, but two C-atoms (C_2 & C_3) are on the plane of the paper and other two (C_1 & C_4) are perpendicular to the plane of the paper. Also all the p-orbitals are along y-axis. C_2(z) is perpendicular to the plane of the paper.

1. **Identity operation (*E*):**

   So, we can write as,

$$\begin{pmatrix} \phi_1' \\ \phi_2' \\ \phi_3' \\ \phi_4' \end{pmatrix} = \begin{pmatrix} \phi_1 \\ \phi_2 \\ \phi_3 \\ \phi_4 \end{pmatrix} \begin{pmatrix} 1 & 0 & 0 & 0 \\ 0 & 1 & 0 & 0 \\ 0 & 0 & 1 & 0 \\ 0 & 0 & 0 & 1 \end{pmatrix}$$

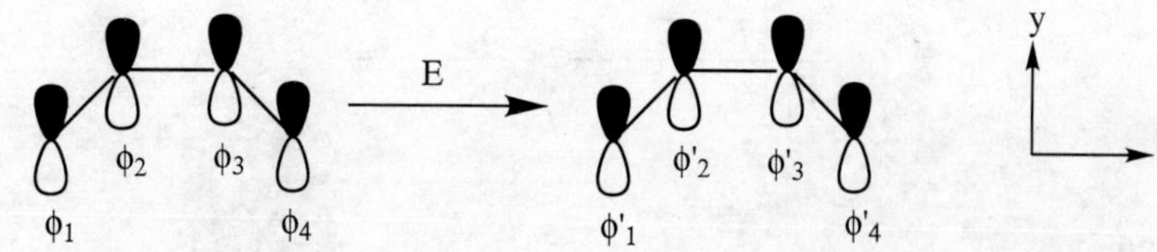

Figure 3.3: Effect of identity operation on four P_y orbitals of 1,3-butadiene.

Thus, the transformation matrix corresponding to the symmetry element E for the unhybridized p_y orbitals of 1,3 butadiene molecule is

$$\begin{pmatrix} 1 & 0 & 0 & 0 \\ 0 & 1 & 0 & 0 \\ 0 & 0 & 1 & 0 \\ 0 & 0 & 0 & 1 \end{pmatrix}$$

Character $\chi_{(E)} = 4$ (Here all the **four** p_y orbital remains unaltered)

2. **C_2(z) operation:**

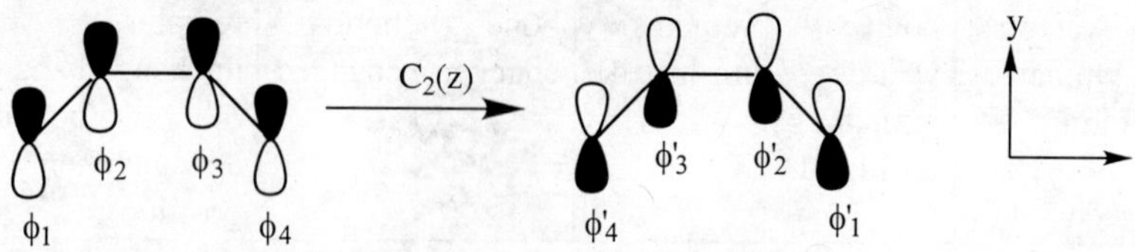

Figure 3.4: Effect of C_2(z) operation on P_y orbitals of 1,3-butadiene

Hence, $\phi_1 = -\phi'_4, \phi_2 = -\phi'_3, \phi_3 = -\phi'_2, \phi_4 = -\phi'_1,$

Therefore, we can write as,

$$\begin{pmatrix} \phi'_1 \\ \phi'_2 \\ \phi'_3 \\ \phi'_4 \end{pmatrix} = \begin{pmatrix} \phi_1 \\ \phi_2 \\ \phi_3 \\ \phi_4 \end{pmatrix} \begin{pmatrix} 0 & 0 & 0 & -1 \\ 0 & 0 & -1 & 0 \\ 0 & -1 & 0 & 0 \\ -1 & 0 & 0 & 0 \end{pmatrix}$$

Hence, the transformation matrix for C_2(z) is

$$\begin{pmatrix} 0 & 0 & 0 & -1 \\ 0 & 0 & -1 & 0 \\ 0 & -1 & 0 & 0 \\ -1 & 0 & 0 & 0 \end{pmatrix}$$

Character $\chi_{(C_2)} = 0$ (Here all the **four** p_y orbitals change their positions)

3. **σ_{xz} operation:**

Here, $\phi_1 = -\phi'_1, \phi_2 = -\phi'_2, \phi_3 = -\phi'_3, \phi_4 = -\phi'_4,$

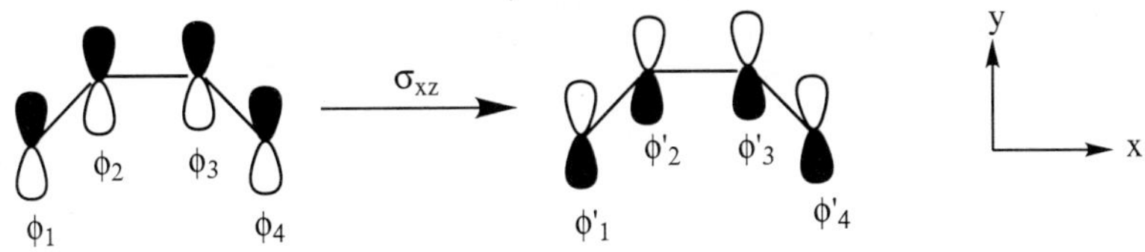

Figure 3.5: Effect of σ_{xz} operation on four P_y orbitals of 1,3-butadiene.

Hence, we can write as,

$$\begin{pmatrix} \phi'_1 \\ \phi'_2 \\ \phi'_3 \\ \phi'_4 \end{pmatrix} = \begin{pmatrix} \phi_1 \\ \phi_2 \\ \phi_3 \\ \phi_4 \end{pmatrix} \begin{pmatrix} -1 & 0 & 0 & 0 \\ 0 & -1 & 0 & 0 \\ 0 & 0 & -1 & 0 \\ 0 & 0 & 0 & -1 \end{pmatrix}$$

Hence, the transformation matrix for σ_{xz} is

$$\begin{pmatrix} -1 & 0 & 0 & 0 \\ 0 & -1 & 0 & 0 \\ 0 & 0 & -1 & 0 \\ 0 & 0 & 0 & -1 \end{pmatrix}$$

Character $\chi_{(\sigma_{xz})} = -4$ (Here all the **four** p_y orbitals change their sign of lobes)

4. σ_{yz} **operation:**

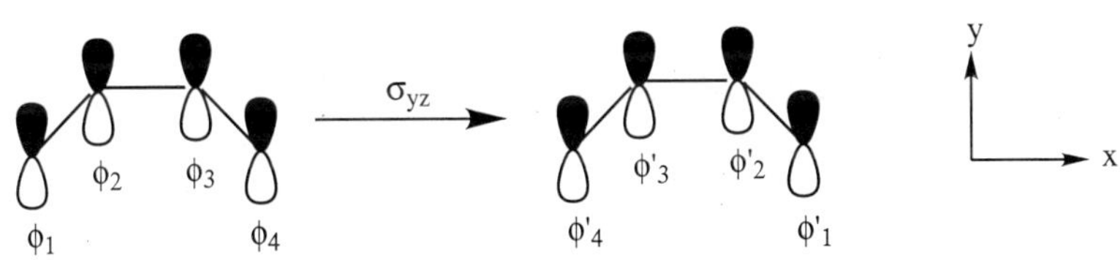

Figure 3.6: Effect of σ_{yz} operation on four P_y orbitals of 1,3-butadiene.

Similar way the transformation matrix is found to be

$$\begin{pmatrix} 0 & 0 & 0 & 1 \\ 0 & 0 & 1 & 0 \\ 0 & 1 & 0 & 0 \\ 1 & 0 & 0 & 0 \end{pmatrix}$$

Character $\chi_{(\sigma_{yz})} = 0$ (Here all the **four** p_y orbitals change their positions)

The reducible representation for 1,3 butadiene, taking all the four unhybridized p_y orbitals, is as follows:

C_{2v}	E	C_2	σ_{xz}	σ_{yz}
Γ_{3N}	4	0	–4	0

In this case we can reach the same table by counting the shifted and unshifted basis as well.

These newly generated reducible representations are of no use in specific problems of quantum mechanics or spectroscopy or bonding or any relevant task. So when we have learnt enough how to get reducible representation for a point group taking some basis, the next crucial step is to find out the hidden irreducible components present in that reducible representation which has direct use in solving any task of our interest.

3.2.2.4 Reducible Representations for Octahedral Geometry Using Bond Vectors as Basis

The reducible representation can be generated by simply counting the number of unshifted and shifted basis or bond vectors and subtracting the later from the former.

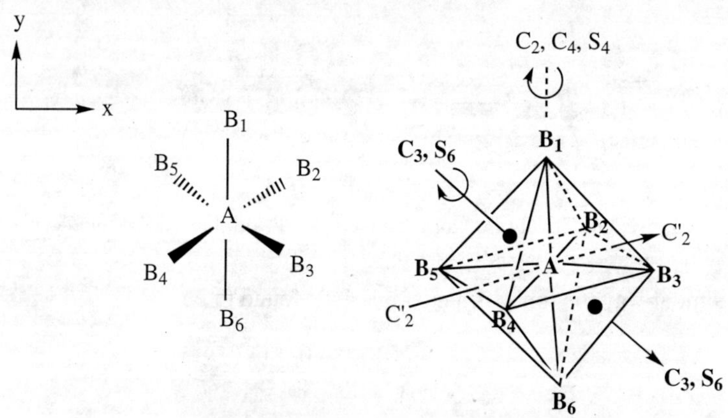

Figure 3.7: Octahedral geometry along with its different symmetry operations

O_h	E	$8C_3$	$6C'_2$	$6C_4$	$3C_2$	i	$6S_4$	$8S_6$	$3\sigma_h$	$6\sigma_d$
Γ_{Red}	**6** (6–0)	**0** (0–0)	**0** (0–0)	**2** (2–0)	**2** (2–0)	**0** (0–0)	**0** (0–0)	**0** (0–0)	**4** (4–0)	**2** (2–0)

3.2.3 Reduction of Reducible Representation to its Component IRs

We have the general formula, derived from GOT, to reduce the reducible representation to its component IRs as:

$$\chi(R) = \sum_j a_j \chi_j(R)$$

From this equation, we get the final form the equation to find the number of a particular irreducible representation (IR) appears in a reducible representation (a_i) as:

$$\mathbf{a_i = (1/h) \sum_R \chi(R) \chi_i(R)}$$

where, χ(R) is the character of the reducible representation and χ_i(R) is the character of the ith IR and h is the order of the group (*i.e.*, number of total operations in the group). This is called *reduction formula* as well.

It can also be written as:

$$\mathbf{a_i = (1/h) \sum_{\text{all classes}} n_R \chi(R) \chi_i(R)}$$

(where, n_R is the number of symmetry operation in the class). We can take different examples to reduce the representation.

Example 1 Let us take one reducible representation of C_{2v} point group as:

C_{2v}	E	C_2	σ_{xz}	σ_{yz}
Γ	3	3	1	1

Here, $h = 4$, $\chi(R) = 3$ for E, and C_2 and $\chi(R) = 1$ for two σ_vs.

From the character table of C_{2v} point group, we see that there are four IRs, A_1, A_2, B_1 and B_2.

For A_1 representation, χ_i(R) = 1 for each operation.

Hence, the number of A_1 occurs in Γ,

$$a_{A_1} = \frac{1}{4}[\underbrace{(3\times1\times1)}_{E}+\underbrace{(3\times1\times1)}_{C_2}+\underbrace{(1\times1\times1)}_{\sigma_{xz}}+\underbrace{(1\times1\times1)}_{\sigma_{yz}}] = 2$$

Here, four terms are there within four parentheses, which corresponds to four symmetry operations of the point group. In each parenthesis, there are three terms; first one designates the character of reducible representation, second onc corresponds to character of irreducible representation corresponding to the symmetry operation and the third one represents the number of symmetry operations present in the class or n_R.

Similarly, number of A_2 occurs in the reducible representation can easily be found using the same formula:

$$\Gamma = 1/4[(3\times1\times1)+(3\times1\times1)+(1\times-1\times1)+(1\times-1\times1)] = 1$$

Number of of B_1 occurs in Γ,

$$\Gamma = 1/4[(3\times1\times1)+(3\times-1\times1)+(1\times1\times1)+(1\times-1\times1)] = 0$$

And similarly number of B_2 occurs is 0. Hence, the reducible representation, Γ reduces to IRs as $2A_1$+A_2.

N.B. in C_{2v}, the two σ's are in two different classes.

Example 2 We can take another example of C_{3v}, where the reducible representation is as follows:

C_{3v}	E	$2C_3$	$3\sigma_v$
Γ	4	1	-2

Number of A_1 occurs in Γ,

$$\Gamma = 1/6[(4\times1\times1)+(1\times1\times2)+(-2\times1\times3)] = 0$$

Number of A_2 occurs in Γ,

$$\Gamma = 1/6[(4\times1\times1)+(1\times1\times2)+(-2\times-1\times3)] = 2$$

And number of E occurs in Γ,

$$\Gamma = 1/6[(4\times2\times1)+(1\times-1\times2)+(-2\times0\times3)] = 1$$

Thus, Γ reduces to $2A_2$+E.

3.2.4 Criteria for Irreducibility

There is a nice way to check whether a given representation is reducible or not. To verify the irreducibility of a representation, a general equation is there:

$$\sum_R n_R[\chi_i(R)]^2 = h;$$

where h order of the point group and n_R is the no. of operations in the class.

Let us take the example of D_3 point group and test whether the representations given in the table below are irreducible or not.

D_3	E	$2C_3$	$3C_2$
Γ_1	1	1	1
Γ_2	1	1	–1
Γ_3	2	–1	0
Γ_x	5	2	–1

For Γ_1, the above eqn can be used

$$\begin{aligned}\Gamma_1 &= \left[\sum n_R[\chi_i(R)]^2\right] \\ &= [1.\chi_i(E)^2 + 2(\chi_i(C_3)^2) + 3(\chi_i(C_2)^2)] \\ &= [1.1^2 + 2.1^2 + 3.1^2] \\ &= [1+2+3] = 6\end{aligned}$$

i.e., it becomes equal to the order of the group (h). Hence, Γ_1 can be concluded as an irreducible representation (IR).
Similarly, for Γ_3,

$$[1.2^2 + 2(-1)^2 + 3(0)^2] = 6 = h$$

Hence, Γ_3 is found to be an IR.
For Γ_x,

$$[1.5^2 + 2.2^2 + 3(-1)^2] = [25+8+3] = 36$$

Since, it is $\neq h$, Hence, this is not an irreducible representation (IR), rather a reducible representation.

EXERCISES

3.1 Determine the matrices for the transformation of six unit vectors along the cartesian coordinate for O_2 molecule (x-axis is along the bond) under C_2 operation and inversion operation.

3.2 Reduce the following representation:

T_d	E	$8C_3$	$3C_2$	$6S_4$	$6\sigma_d$
Γ	9	3	1	3	3

3.3 Reduce the following reducible representation:

C_{3v}	E	$2C_3$	$3\sigma_v$
Γ	15	0	3

3.4 Show the reducibility character of the following representation:

D_{3h}	E	$2C_3$	$3C_2$	σ_h	$2S_3$	$3\sigma_v$
Γ	12	0	–2	4	–2	2

3.5 Using p_x, p_y and p_z orbitals as basis on the central atom for a tetrahedral point group molecule, find the reducible representation.

Chapter 4

Great Orthogonality Theorem and its Consequence

4.1 Introduction

All the properties of a group and their characters are very important in solving problems related to chemical bonding or molecular dynamics *etc.*, can be derived from one basic theorem that takes into account the relations among matrix elements of the irreducible representations corresponding to their reducible representations. The theorem is known as **Great Orthogonality Theorem** or in short termed as **GOT**. Here, we are not entering into the complicacy of derivation for this theorem, but the expression of the theorem is:

$$\sum[\Gamma_i(R)_{mn}][\Gamma_j(R)_{m'n'}]^* = \frac{h}{\sqrt{d_i d_j}}\delta_{ij}\delta_{mm'}\delta_{nn'} \tag{4.1}$$

Where,

h is order of the group, *i.e.*, number of operations in a group.

d_i is dimension of ith IR representation *i.e.*, the character under the identity operation, E.

R represents symmetry operations.

$\Gamma_i(R)_{mn}$ is the element of mth row and n^{th} column of matrix corresponding to the symmetry operation, R in i^{th} IR, Γ_i.

$[\Gamma_j(R)_{m'n'}]^*$ is similar to the previous one, but for j^{th} IR and the * indicates that, if the character is imaginary, then corresponding complex conjugate is considered.

$\delta's$ are kronecker delta and can have values either 0 or 1.

4.2 General Rules Derived from GOT

There are several general rules derived from GOT which have several applications. These are discussed one by one taking examples. It is very much helpful, particularly, during construction or verification of character table. The rules are:

1. *The sum of the squares of the dimensions of the IR's of a point group is equal to the order of the group, i.e.,*

$$\sum d_i^2 = h$$

Where d_i's are positive integer and h is order of a point group. Remember that the character of each IR under the symmetry element, E gives the dimension of the IR. Hence, the above rule can be rewritten as:

$$\sum[\chi_i(E)]^2 = h$$

Let us recall the partial T_d point group character table, where total number of symmetry operations are 24, *i.e.*, h =24. There are five IRs ($\Gamma's$) present in this point group.

Table 4.1: Partial character table of T_d point group.

T_d	E	$8C_3$	$3C_2$	$6S_4$	$6\sigma_d$	*Dimension* (d_i)	$(d_i)^2$
$\Gamma_1 = A_1$	1	1	1	1	1	$d_1 = 1$	1
$\Gamma_2 = A_2$	1	1	1	-1	-1	$d_2 = 1$	1
$\Gamma_3 = E$	2	-1	2	0	0	$d_3 = 2$	4
$\Gamma_4 = T_1$	3	0	-1	1	-1	$d_4 = 3$	9
$\Gamma_5 = T_2$	3	0	-1	-1	1	$d_5 = 3$	9

From the last column of the Table 4.1, it is evident that the sum of the squares of the dimensions of all irreducible representations (IRs) is 24.

$$[1^2+1^2+2^2+3^2+3^2 = 1+1+4+9+9 = 24].$$

2. *The sum of the squares of the characters of matrices corresponding to all R in the ith IR equals to the order of the point group, i.e.,*

$$\sum[\chi_i(R)]^2 = h$$

From the above character table, if we consider any IR (say Γ_4 or T_1), then calculating $\sum[\chi_i(R)]^2$ we get exactly equal to the order of the point group, *i.e.*, h.

$$[(3)^2+8(0)^2+3(-1)^2+6(1)^2+6(-1)^2 = 24].$$

3. *The vectors or functions whose components are the characters of two different IR's, must be orthogonal to each other.*

Two representative IRs (say A_2 and T_2) are taken and have to prove that $[A_1 \times T_2] = 0$. If we now, multiply characters corresponding to each symmetry operation for two IRs and sum up, we get,

$$[(1\times3)+8(1\times0)+3(1\times-1)+6(-1\times-1)+6(-1\times1)] = 0$$

Here, one point must be noted that the number of symmetry operations in a class must have to consider and needs to multiply accordingly.

4. *In a given reducible or irreducible representation, the characters of matrices corresponding to operations, belonging to the same class, are identical, i.e., in a class all matrices have same trace or χ- value.*
   Here, in T_d point group, for example, for C_3 element, there are four C_3^1 operations and four C_3^2 operations, but all the operations produce same character for a particular IR and hence they are clubbed together. For T_1 IR, all eight operations corresponding to C_3 element, produce same 'zero' character. Similarly, for all three C_2 operations have same character, –1.
5. *The number of IRs present in a point group, equals to the number of classes in that point group.*
   From the character of T_d point group, it can be seen that there are five classes (columns wise) and five IRs as well (row wise).
6. *One dimensional representation, in which all the characters are listed as +1, called totally symmetric representation, will always be there in a point group.*

   In this point group, it is the IR, A_1 which plays the role of totally symmetric representation.
7. *The number of times any IR appears (a_i) as a component in any reducible representation is given by the equation:*

$$a_i = \frac{1}{h}\left[\sum_{classes} n_R \chi(R) \chi'_i(R)\right]$$

   Where, n_R is the number of symmetry operations in a class, $\chi(R)$ is the character of reducible representation corresponding to R symmetry operation and $\chi'_i(R)$ is the the character of the ith irreducible representation corresponding to the symmetry operation, R. The above formula is known as reduction formula.

8. *Characters of an IR of a point group are independent of the choice of basis.*
   Basis are the functions, vectors or orbitals or axis on which symmetry operations are performed to generate reducible representations. There are several special basis sets for representation and some of these basis sets are:
   - Unit vectors x, y, z along Cartesian coordinates or translational vectors T_x, T_y, T_z.
   - Rotational vectors R_x, R_y, R_z about their respective axis.
   - Vectors directed along the various bonds of a molecule.
   - Wave functions of a molecule, *etc.*

4.3 Illustrations

The rules stated above, can be illustrated taking few examples. Character tables corresponding to different point groups can be constructed using those rules as well.

4.3.1 C_{2v} Point Group

1. It consists of symmetry operations E, $C_2(z)$, σ_{xz}, σ_{yz} and each operation belongs to a separate class. Hence, there are four classes in C_{2v} point group and hence from the rules obtained from GOT, the total number of IRs will also be four (say, $\Gamma_1, \Gamma_2, \Gamma_3, \Gamma_4$).
2. Since, there must be one totally symmetric IR. Let us consider, Γ_1 be the totally symmetric IR.
3. Let, the dimension of the four IRs be d_1, d_2, d_3 and d_4. According to rule, $d_1^2 + d_2^2 + d_3^2 + d_4^2 = h = 4$. Since, dimension of any IR can not be zero or negative, hence, the condition can only be satisfied if and only if all d's have values +1. Therefore, all the four IRs are of dimension one. We can summarize the information gathered so far in a table as:

C_{2v}	E	$C_2(z)$	σ_{xz}	σ_{yz}
Γ_1	1	1	1	1
Γ_2	1			
Γ_3	1			
Γ_4	1			

4. Consider that the characters of Γ_2 be 1, χ_1, χ_2 and χ_3. We have already, $\Sigma[\chi_i(R)]^2 = 4$. Hence, $(1)^2 + (\chi_1)^2 + (\chi_2)^2 + (\chi_3)^2 = 4$ Or, $(\chi_1)^2 + (\chi_2)^2 + (\chi_3)^2 = 3$

   This is only possible if $\chi_i = \pm 1$ (i = 1, 2, 3). Here, we have considered Γ_2, but this is true for any IR of this group. Hence, the table can further be extended as:

C_{2v}	E	$C_2(z)$	σ_{xz}	σ_{yz}
Γ_1	1	1	1	1
Γ_2	1	± 1	± 1	± 1
Γ_3	1	± 1	± 1	± 1
Γ_4	1	± 1	± 1	± 1

All other IRs must be such that the above equation is satisfied. This will be valid only if $\chi(R) = 1$. According to rule 3, all other IRs $\Gamma_2, \Gamma_3, \Gamma_4$ must be orthogonal to Γ_1. Thus, to fulfill this criteria, those $\Gamma_2, \Gamma_3, \Gamma_4$ representations will have characters as +1 for two symmetry operations and characters as –1 for other two symmetry operations. Thus, the four IRs will have the following characters shown in the final table (Table 4.2) Orthogonality of these four representations can be easily verified.

For example, let us take the characters of Γ_3 and Γ_4 and add after multiplication of corresponding characters. $(1\times) + (-1\times -1) + (-1\times 1) + (1\times -1) = 0$.

For, Γ_2 and Γ_3, similar exercise gives, $(1\times 1) + (1\times -1) + (-1\times -1) + (-1\times 1) = 0$; Hence, it can be concluded that C_{2v} has four different IRs with characters either +1 or –1 and each IR is orthogonal to every other IRs.

Table 4.2: Characters of C_{2v} point group.

C_{2v}	E	$C_2(z)$	σ_{xz}	σ_{yz}
Γ_1	1	1	1	1
Γ_2	1	1	–1	– 1
Γ_3	1	–1	–1	1
Γ_4	1	–1	1	–1

4.3.2 D_3 Point Group

Let us take an example, of $[Co(en)_3]^{3+}$. It has E, $2C_3$ and $3C_2$ symmetry operations. Therefore, it has total six operations and hence, the order of the group, h = 6. Since,there are three classes in this group, hence, there would be total three IRs in the group (rule 5). Let the dimension of the IRs be d_1, d_2, d_3 and according to rule 1:

$$d_1^2 + d_2^2 + d_3^2 = h = 6$$

After careful inspection, it is found that only possible values of these dimensions are: $d_1 = 1$, $d_2 = 1$ and $d_3 = 2$. It is known that, there must be one totally symmetric IR in which the characters for all symmetry operations in all classes are +1 only. Thus, we can start tabulating as follows:

D_3	E	$2C_3$	$3C_2$
Γ_1	1	1	1

It is seen that the characters of Γ_1 satisfies the relation,

$$\sum n_R[\chi_i(R)]^2 = h$$

$$i.e., 1 \times (1)^2 + 2 \times (1)^2 + 3 \times (1)^2 = 6$$

Now, we look for a vector in 6-dimensional space (as h =6) whose components must be equal to ±1 and orthogonal to Γ_1 as well. Components of vectors or characters must contain three (+1) and three (–1). Since, $\chi(E)$ indicates the dimension of an IR and always must be a positive integer. This is also known that all the symmetry operations in a class must have same characters. Considering all these constraints, along with orthogonal relationship between Γ_1 and Γ_2, $[(1 \times 1) + 2 \times (1 \times 1) + 3 \times (1 \times -1) = 0]$ the following table can be constructed. The third IR (Γ_3) will be of dimension 2, *i.e.*, d_3 = 2 or, χ(E) = 2 to fulfil the criteria

D_3	E	$2C_3$	$3C_2$
Γ_1	1	1	1
Γ_2	1	1	–1

of $\sum[\chi_i(E)]^2 = h$. Let the character for C_3 is $\chi(C_3)$ and the character for C_2 is $\chi(C_2)$. Incorporating Γ_3, the table can be extended as given below. At this point, finding the two characters for Γ_3 has become very easy. Using rule 2 and rule 3, the characters can easily be found. First, if we try using rule 2, *i.e.*, $\sum[\chi_i(R)]^2 = h$, which gives an

D_3	E	$2C_3$	$3C_2$
Γ_1	1	1	1
Γ_2	1	1	–1
Γ_3	2	$\chi(C_3)$	$\chi(C_2)$

idea of characters, $\chi(C_3)$ and $\chi(C_2)$ to be ± 1. Making use of the rule of orthogonality for Γ_1 and Γ_3, we get

$$(1\times 2)+2\times(1\times\chi(C_3))+3\times(1\times\chi(C_2))=0$$

Or,

$$2+2\chi(C_3)+3\chi(C_2)=0\ldots\ldots(i)$$

Again, orthogonality check for Γ_2 and Γ_3, we get,

$$2+2\chi(C_3)-3\chi(C_2)=0\ldots\ldots(ii)$$

Adding these two equations,

$$4+4\chi(C_3)=0,\ \textit{i.e.}\ \chi(C_3)=-1$$

And subtracting the two equations,

$$6\chi(C_2)=0,\ \textit{i.e.}\ \chi(C_2)=0$$

Therefore, the complete set of characters of all three IRs are now given in the Table 4.3.

Finally, the correctness of Γ_3 can be checked further by making use of the rule 2, *i.e.*,

$$[2^2+2\times(-1)^2+3\times(0)^2]=6$$

Table 4.3: Characters of D_3 point group.

D_3	E	$2C_3$	$3C_2$
Γ_1	1	1	1
Γ_2	1	1	–1
Γ_3	2	–1	0

EXERCISES

4.1 State the general rules derived from Great Orthogonality Theorem.

4.2 Using the d-orbitals as basis for a tetrahedral molecule, first construct reducible representation and then reduce it to its irreducible components and also show that these components are degenerate.

4.3 Using GOT, find the components of the following reducible representation:

D_{3h}	E	$2C_3$	$3C_2$	σ_h	$2S_3$	$3\sigma_v$
Γ	12	0	–2	4	–2	2

4.4 For point group C_{3v}, one reducible representation is [Γ= 6,0,0] corresponding to the three symmetry elements. Show that $\Gamma = A_1 + A_2 + E$.

4.5 Using the general rules derived from GOT, find the missing characters in the following table:

C_{3v}	E	$2C_3$	σ_v
Γ_1	1	1	–
Γ_1	1	1	–
Γ_1	–	–1	0

Chapter 5

Character Table and Direct Product

5.1 Introduction

Character tables are quantitative representation and characteristics of point groups and very useful in dealing with practical problems like, molecular spectroscopy or other related fields. We have seen that how GOT can be utilized to find the total number of IRs of a point group as well as characters of IRs.

5.2 Anatomy of Character Table

The character table has the following lay out and it is divided into six sections, *a* to *f* as shown below:

Table 5.1: Different parts of a character table.

a	b	e	f
c	d		

1. **Section (a)** gives the Schoenflies symbol for the point group, *i.e.* C_{2v}, C_{2h}, D_{2h} *etc.* which provides the very basic knowledge about a point group.
2. **Section (b)** lists the symmetry operations of the point group in class wise manner.
3. **Section (c)** lists the conventional symbols/notations for various IRs, *i.e.* Mulliken's symbols for IRs.
4. **Section (d)** gives the characters of all IRs for a symmetry operation in each class of symmetry operations.
5. **Section (e)** lists the basis sets of representation for each of the IR. These may be x, y, z axes; T_x, T_y, T_z translational vectors; R_x, R_y, R_z rotational vectors; sets of orbitals; bond vectors, *etc.*

6. **Section (f)** shows that how the binary/ternary functions of x, y, z form the basis for representations for some of the IRs. Binary and ternary combinations can be x^2, y^2, z^2; xy, xz, yz; $(x^2 - y^2), (x^2 + y^2), x^3$ and so on. These are important in dealing with symmetries of different orbitals and Raman spectroscopy.
7. **Section (c)** of the character table gives the Mulliken's symbol/notations, *i.e.*, names for the IRs. The symbols are assigned as per the following convention:
   - All one dimensional IRs (characters under E operation) are labeled as A or B.
   - All two dimensional IRs are labeled as E.
   - All three dimensional representations are labeled as T or F. Usually T is used in electronic transitions and F is used in vibrational related problems. The higher order representations are represented by G (four dimensional), H (five dimensional) and so on.
   - For point groups $C_{\alpha v}$ and $D_{\alpha h}$ (linear molecules), a different notation is used. One dimensional IR is represented by the symbol Σ, two dimensional IR is represented by Π and three dimensional IR is represented by Δ.
8. In addition to these general symbols, most Mulliken's symbols have different subscripts and/or superscripts which follow the following conventions.
   - An one dimensional IR is given symbol 'A' or 'B' if the basis is symmetric (*i.e.*, $\chi = +1$) or antisymmetric (*i.e.*, $\chi = -1$) with respect to rotation about the highest order proper axis (Principal axis) or improper axis, respectively. For C_1, C_s, C_i (which do not have any C_n axis), symbol 'A' is used for one dimensional IR.
   - Subscript '1' or '2' is used with A or B if basis is symmetric and antisymmetric with respect to the rotation about those C_2 axis which are perpendicular to principal C_n axis ($C_2 \perp C_n$), respectively. If such C_2 axis is lacking, σ_v is considered for the same purpose. For D_2 and D_{2h} point groups where there are three C_2 axes and these axes fall in different classes, the IR in which χ for all C_2 axes are +1 is labeled as 'A', and in which χ for all C_2 axes are not +1 is labeled as 'B'. For D_{2d} point group, the character of S_{2n} determines the symbol.
   - For linear molecules, the symmetric or antisymmetric in above rule are indicated by + or – sign in superscript, *i.e.*, Σ^+ or Σ^- instead of A_1 or A_2.
   - Subscript g or u are used respectively to IRs if symmetric or antisymmetric with respect to centre of inversion, *i. e.g.*, $A_g, B_g, A_{1u}, B_{1u}, etc.$
   - Superscripts single prime ($\prime$) or double prime ($\prime\prime$) are used to those IRs which are symmetric or antisymmetric with respect to σ_h respectively. These last two rules are equally applicable to one-, two-, three-dimensional IRs.

5.3 Construction of Character Table

Already we have seen how to generate the characters of a character table, in the previous chapter. Here again, along with the examples illustrated in the previous chapter, we will extend the discussion to construct the character table for C_{4v} point group. According to the rules obtained from GOT, total number of operations in this point group are eight and these symmetry operations have been divided into five classes. Hence, there will be only five IRs $(\Gamma_1, \Gamma_2, \Gamma_3, \Gamma_4, \Gamma_5)$ with dimensions d_1, d_2, d_3, d_4, and d_5. From the rule 1 of GOT, we know that the sum of the squares of the dimensions is equal to the order of the group, *i.e.*, $d_1^2 + d_2^2 + d_3^2 + d_4^2 + d_5^2 = h = 8$.

This condition will only be satisfied when $d_1 = d_2 = d_3 = d_4 = 1$ and $d_5 = 2$. As we know that the dimension of IR equals to character under identity operation, E, hence, these are the five characters values under the column designated for operation E. We also know that there must be a totally symmetric IR for any point group. Therefore, the partial character table for C_{4v} point group looks as:

Table 5.2: Partial characters of C_{4v} point group.

C_{4v}	E	$2C_4$	C_2	$2\sigma_v$	$2\sigma_d$
Γ_1	1	1	1	1	1
Γ_2	1				
Γ_3	1				
Γ_4	1				
Γ_5	2				

Let the remaining characters for Γ_2 be $\chi(C_4), \chi(C_2), \chi(\sigma_v)$ and $\chi(\sigma_d)$. Character for E is already known as +1.

Applying orthogonality rule for Γ_1 and Γ_2 we find,

$$[1 \times 1 \times \chi(E) + 2 \times 1 \times \chi(C_4) + 1 \times 1 \times \chi(C_2) + 2 \times 1 \times \chi(\sigma_v) + 2 \times 1 \times \chi(\sigma_d)] = 0$$

Since, already known that $\Gamma_1, \Gamma_2, \Gamma_3, \Gamma_4$ are one-dimensional, following the rule 2, the values of $\chi(C_4), \chi(C_2), \chi(\sigma_v)$ and σ_d for all the four IRs would be either +1 or –1 only. Hence, at this stage, the form of the character table would be:

C_{4v}	E	$2C_4$	C_2	$2\sigma_v$	$2\sigma_d$
Γ_1	1	1	1	1	1
Γ_2	1	±1	±1	±1	±1
Γ_3	1	±1	±1	±1	±1
Γ_4	1	±1	±1	±1	±1
Γ_5	2	$\pm\chi(C_4)$	$\pm\chi(C_2)$	$\pm\chi(\sigma_v)$	$\pm\chi(\sigma_d)$

From GOT rule, we also know that in a class, the squares of the characters for symmetry operation when multiplied by number of symmetry operations in that class and summed over all IRs will be equal to the order of the group.

Let us take C_2 class and apply this rule (down the column):

$$[(1)^2 + (\pm1)^2 + (\pm1)^2 + (\pm1)^2 + (\pm\chi C_2)^2] = 8$$

This is possible only when $\chi(C_2)$ in Γ_5 is either +2 or –2.

The class corresponding to C_4 is taken and follow the similar procedure, *i.e.*,

$$[2 \times (1)^2 + 2 \times (\pm1)^2 + 2 \times (\pm1)^2 + 2 \times (\pm1)^2 + 2 \times (\pm\chi C_4)^2] = 8$$

$\chi(C_4)$ is found to be 0 in Γ_5.

Similarly, in Γ_5, $\chi(\sigma_v) = 0$ and $\chi(\sigma_d) = 0$ can be obtained .

The table now looks as shown in Table 5.3.

Table 5.3: Partial characters of C_{4v} point group.

C_{4v}	E	$2C_4$	C_2	$2\sigma_v$	$2\sigma_d$
Γ_1	1	1	1	1	1
Γ_2	1	±1	±1	±1	±1
Γ_3	1	±1	±1	±1	±1
Γ_4	1	±1	±1	±1	±1
Γ_5	2	0	±2	0	0

Further, we can test the orthogonality of Γ_1 and Γ_5:

$$[1\times1\times2+2\times1\times0+1\times1\times(\pm2)+2\times1\times0+2\times1\times0] \text{ must be } 0$$

In order to fulfil this criteria, $\chi(C_2)$ must be –2 in Γ_5.

Putting the value of $\chi(C_2)$, we can test the orthogonality of Γ_4 and Γ_5:

$$[1\times1\times2+2\times(\pm1)\times0+1\times(\pm1)\times(-2)+2\times(\pm1)\times0+2\times(\pm1)\times0] \text{ must be equal to } 0$$

In order to fulfil this criteria, $\chi(C_2)$ must be +1 in Γ_4. Hence, the table becomes as shown in Table 5.4.

Table 5.4: Partial characters of C_{4v} point group.

C_{4v}	E	$2C_4$	C_2	$2\sigma_v$	$2\sigma_d$
Γ_1	1	1	1	1	1
Γ_2	1	±1	±1	±1	±1
Γ_3	1	±1	±1	±1	±1
Γ_4	1	±1	1	±1	±1
Γ_5	2	0	-2	0	0

Now, $\Gamma_1,\Gamma_2,\Gamma_3,\Gamma_4$ are one-dimensional IRs and their characters are ±1. Since, there are eight operations in the group, four of these must have character +1 and other four must have character –1 so that orthogonality condition is obeyed by these four IRs. As we have derived $\chi(C_2)$ = +1 for Γ_4, similarly, we can derive $\chi(C_2)$ for Γ_2,Γ_3 by making use of their orthogonality relation with Γ_5. Γ_2 and Γ_5:

$[1\times1\times2+2\times(\pm1)\times0+1\times(\pm1)\times(-2)+2\times(\pm1)\times0+2\times(\pm1)\times0]$ must be equal to 0.

To satisfy the equation, the product under C_2 must be of negative sign and the character of C_2 under Γ_5 is already –2. Therefore, $\chi(C_2)$ must be +1 in Γ_2.

Similarly, same exercise is performed for $\chi(C_2)$ in Γ_3 and it is found to be +1. The remaining characters may be written as +1 or –1 in $\Gamma_1,\Gamma_2,\Gamma_3,\Gamma_4$ so that there would be four +1 and four –1 characters. Therefore, the final table can be written as shwon in Table 5.5.

Table 5.5: Final characters of C_{4v} point group.

C_{4v}	E	$2C_4$	C_2	$2\sigma_v$	$2\sigma_d$
Γ_1	1	1	1	1	1
Γ_2	1	1	1	–1	–1
Γ_3	1	–1	1	1	–1
Γ_4	1	–1	1	1	–1
Γ_5	2	0	–2	0	0

5.4 Direct Product

In mathematics, specifically in group theory, the direct product is an operation that involves two different groups (say, A and B) and constructs a new group which usually denoted by $A \times B$. Last column of any character table lists the symmetries of direct products between linear vectors or matrices. The direct product can be taken among any numbers of irreducible representations (IR) in a character table of any point group. The use of direct product representations is very important in quantum mechanics, particularly in getting selection rules for spectral transitions *etc.* Direct product between two or more irreducible representations can be written as

$$\Gamma_{\alpha\beta\gamma} = \Gamma_\alpha . \Gamma_\beta . \Gamma_\gamma$$

To generate direct product, we simply multiply together the characters of the component representation, operation by operation. The resulting representation may or may not be reducible. Some important characteristics of direct product representation are as follows:

1. *If all the combined IRs are nondegenerate, the product will also be a nondegenerate representation.*

   This statement can be illustrated taking partial C_{4v} character table as shown in Table 5.6. When all the characters under the IR are ± 1, it is called *nondegenerate* representation and if any character of IR is greater than 1, is called *degenerate* representation.

   Table 5.6: Illustration of direct product in C_{4v} point group.

C_{4v}	E	$2C_4$	C_2	$2\sigma_v$	$2\sigma_d$
A_1	1	1	1	1	1
A_2	1	1	1	–1	–1
B_1	1	–1	1	1	–1
B_2	1	–1	1	–1	1
E	2	0	–2	0	0

   If we consider two nondegenerate IRs, B_1 and B_2, the direct product, $B_1.B_2$ results A_2 which is a nondegenerate representation, as shown below in Table 5.7.

2. *The product between a nondegenrate representation and a degenerate representation is always a degenerate representation.*

Table 5.7: Illustration of direct product in C_{4v} point group.

B_1	1	–1	1	1	–1
B_2	1	–1	1	–1	1
$B_1.B_2$	1	1	1	–1	–1

For example, from the C_{4v} character table (Table 5.6), $B_2.E = E$ or $A_1.E = E$ and E is a degenerate representation.

3. *The direct product of any representation with the totally symmetric representation results that representation itself. e.g.,* $A_1.E = E$; $A_1.B_1 = B_1$ *etc.*
4. *The direct product of two or more degenerate representations is a reducible representation.*

   For example, the result of the direct product, $E.E$ does not match with the existing representation of the group. From reducibility test it is found that this is a reducible representation. Using reduction formula, after reduction, it produces its component representation as, $A_1 + A_2 + B_1 + B_2$.
5. *The direct product of a nondegenerate IR with itself always results totally symmetric representation.*

   Since, the characters corresponding to any nondegenerate IR are all either +1 or –1, hence, the square term of (± 1) always produces +1 which corresponds to totally symmetric representation, or in this case, A_1. *e.g.,* $A_1.A_1 = A_1$, $A_2.A_2 = A_1$, $B_1.B_1 = A_1$ *etc.*
6. *The direct product of a degenerate IR with itself must contain totally symmetric representation as one of the components.*

   We have already seen in the forth point that $E.E = A_1 + A_2 + B_1 + B_2$. Here, the totally symmetric representation, A_1 is one of the components. This can be verified taking any other degenerate representation from other point groups.

In addition to the points above, we can make some generalizations about the Mulliken symbols of direct products between representations. We know that g and u subscript indicate symmetry and antisymmetry with respect to inversion. Similarly prime ($'$) and double prime ($''$), used as superscripts with Mulliken symbols, indicate the symmetry and antisymmetry with respect to horizontal plane. The direct products of representations with either of these symbols follow the general rule as follow:

$$(g).(g) = (g); \quad (u).(u) = (g); \quad (g).(u) = (u)$$

and

$$(').(') = ('); \quad ('').('') = (''); \quad (').('') = ('')$$

These results have consequences for developing selection rules in spectroscopy, as well as other applications in quantum mechanics.

EXERCISES

5.1 Write the important characteristics about direct product representation.

5.2 From the table below, find the direct products of the following and also reduce to their corresponding components where applicable.
(i) $\Gamma_1 \times \Gamma_2$ (ii) $\Gamma_2 \times \Gamma_2$ (iii) $\Gamma_2 \times \Gamma_3$ (iv) $\Gamma_3 \times \Gamma_3$ (v) $\Gamma_5 \times \Gamma_5$ (vi) $\Gamma_4 \times \Gamma_5$

T_d	E	$8C_3$	$3C_2$	$6S_4$	$6\sigma_d$
Γ_1	1	1	1	1	1
Γ_2	1	1	1	–1	–1
Γ_3	2	–1	2	0	0
Γ_4	3	0	–1	1	–1
Γ_5	3	10	–1	–1	1

5.3 From the above table, try to name each Γ according to the Mulliken symbol.

5.4 Try to construct the character table for C_{2h} point group using the general rules derived from GOT.

5.5 What general information is obtained from the following Mulliken symbols for various character table:

$$A_1, A_2', A_u, B_{1g}, E_u, T_{1u}, \Sigma_g^+, \Sigma_u^-$$

Chapter 6

Determination of Symmetries of Vibrations of a Molecule

6.1 Introduction

The knowledge of symmetry of vibrations of a molecule is very important. Knowledge of symmetry gives idea whether a vibrational mode would be Infrared or Raman active and would be visible in spectra or not. It gives freedom to conclude whether a particular vibrational mode would interact with other vibration to elucidate experimentally observed spectra. In this chapter we will learn to find out the symmetry nature of vibrational modes of molecules.

We have already learnt to find character for different symmetry operations as well as to construct reducible representation for a molecule taking suitable basis. Group theory can be an enormous help in deciding the IR/Raman activity of different molecular vibrations.

We know that for a molecule with N-atoms, the number of vibrations are 3N-5 (for linear molecule) and 3N-6 (for nonlinear molecule). Using IR and Raman data all those vibrations can be assigned systematically. Let us start our discussion with few example.

6.2 H_2O Molecule (C_{2v} Point Group)

In H_2O molecule, each atom is taken as origin and corresponding Cartesian coordinates are assumed (Figure 6.1). Considering these Cartesian coordinates as basis, the reducible representation can be constructed and is shown in Table 6.1.

Using the character table of C_{2v} and applying the reduction formula,

$$a_i = \frac{1}{h}\left[\sum_{all\ classes} n_R.\chi(R).\chi_i(R)\right]$$

we get the relative contribution of the appropriate irreducible representations present in the point group.

(Here, a_i is the number of ith IR appears in the reducible representation, n_R is the number of symmetry operations present in a class, $\chi(R)$ is the character of reducible representation under different symmetry operations and $\chi_i(R)$ is the character of i-th IR under different symmetry operations.)

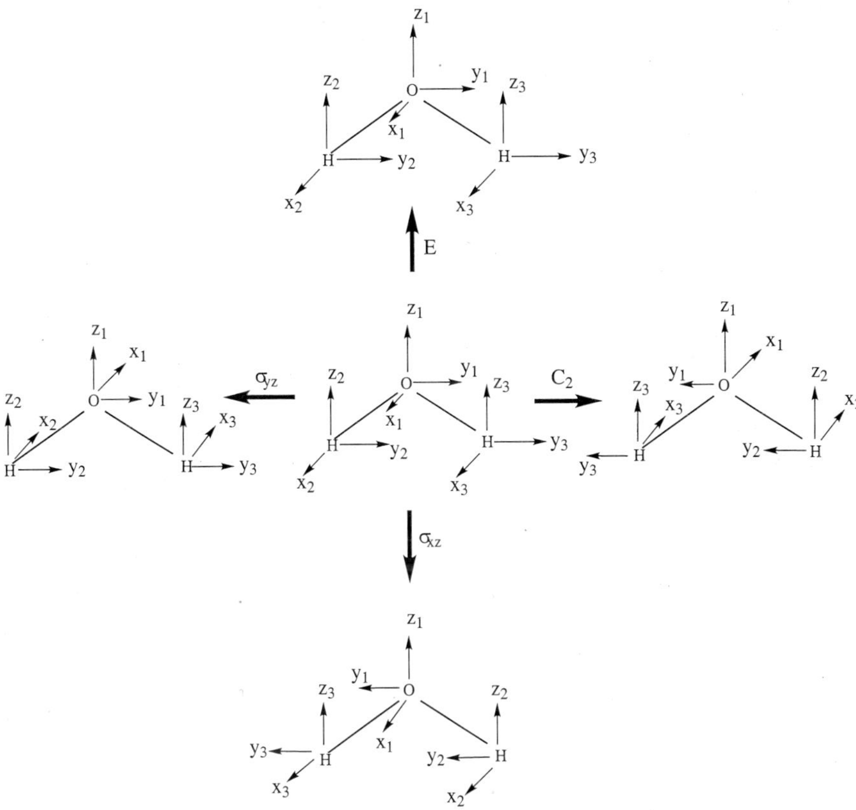

Figure 6.1: Effect of different symmetry operations on vectors along Cartesian coordinates.

Table 6.1: Reducible representation of C_{2v} point group.

C_{2v}	E	C_2	σ_{xz}	σ_{yz}
Γ_{red}	9	–1	1	3

$$\text{Number of } A_1 : \frac{1}{4}[(1\times 9)+(1\times -1)+(1\times 1)+(1\times 3)] = 3$$

$$\text{Number of } A_2 : \frac{1}{4}[(1\times 9)+(1\times -1)+(-1\times 1)+(-1\times 3)] = 1$$

$$\text{Number of } B_1 : \frac{1}{4}[(1\times 9)+(-1\times -1)+(1\times 1)+(-1\times 3)] = 2$$

$$\text{Number of } B_2 : \frac{1}{4}[(1\times 9)+(-1\times -1)+(-1\times 1)+(1\times 3)] = 3$$

Therefore,

$$\Gamma_{red} = 3A_1 + A_2 + 2B_1 + 3B_2$$

These are the symmetry species of total nine possible molecular movements. To get symmetries due to only vibrational movements, we need to remove the symmetries of translational and rotational movements from Γ_{red} or Γ_{total}. The translations must be affected by the group operations in the same way corresponding to x, y and z directions (*i.e.*, B_1, B_2 and A_1 respectively, see C_{2v} character table). Similarly, the three rotations belong to A_2, B_1 and B_2 symmetry species (as specified by R_x, R_y, and R_z respectively, in the character table).

Hence, it can be concluded that symmetry components due to different vibrational modes are:
Γ_{Vib} = [(Total degrees of freedom, *i.e.,* Symmetries of all molecular movements) - (Symmetries of translations + Symmetries of rotations)]

$$\Gamma_{Vib} = (3A_1 + A_2 + 2B_1 + 3B_2) - [(A_1 + B_1 + B_2) + (A_2 + B_1 + B_2)]$$

Or,

$$\Gamma_{Vib} = 2A_1 + B_2$$

It is seen that among the three vibrations of water molecule, two vibrations are of A_1 symmetry and one is of B_2 symmetry.

At this point, it will be very interesting to find out the nature of symmetry (whether A_1 or B_2) conforming to a specific vibration (symmetric stretching, asymmetric stretching or bending). For that let us take appropriate vectors as basis to describe a particular mode of vibration and see the effect of different symmetry operations of C_{2v} point group.

First, we consider two vectors along the two O–H bonds and also assumed that the two bonds can move (whether stretching or compression) independently in plane.

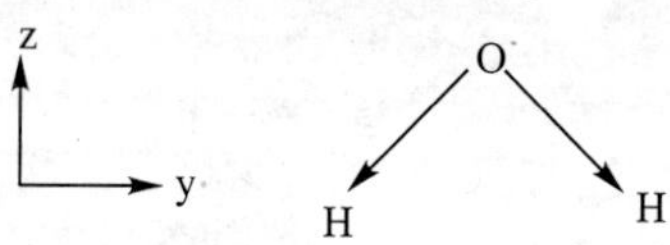

Using these O–H stretching vectors as basis, the reducible representation can be constructed as follows:

C_{2v}	E	C_2	σ_{xz}	σ_{yz}
$\Gamma_{\text{O-H}}$	2	0	0	2

Using reduction formula, $\Gamma_{\text{O–H}}$ is found to be reduced to $A_1 + B_2$ irreducible combination. It indicates that the two O-H bonds can move in plane in two ways: one following A_1 symmetry and the other one following B_2 symmetry. These symmetry labels basically describes the stretching of O–H bonds; symmetric stretching and antisymmetric stretching respectively, shown below.

6.2.1 Symmetric Stretching Mode

We need to go into further detail to designate specific symmetry for specific stretching vibrational mode. The expansion of bonds is considered as positive phase and

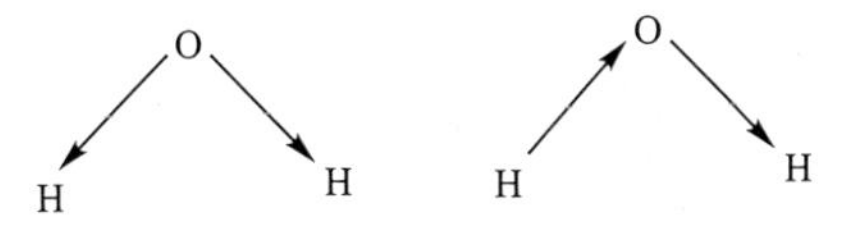

contraction as negative phase. The effect of different symmetry operations on these two phases is now observed. If after the operation the phase changes, then the contribution will be –1 and if remains same, the contribution will be +1.

It is evident from the Figure 6.2 that under all the the symmetry operations, the symmetric stretching mode of vibration does not change its phase *i.e.*, positive phase remains positive after symmetry operations (We could have started with negative phase and would see that negative phase remains negative after all the symmetry operations) and produces the characters (Γ_1) same as A_1 irreducible representation in the character table of C_{2v} (Table 6.2).

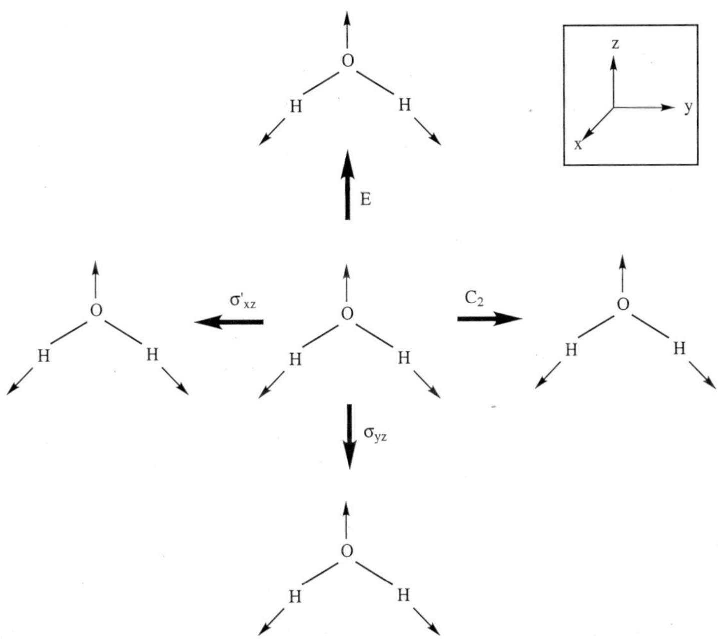

Figure 6.2: Effect of symmetry operations on symmetric stretching mode of vibration.

Table 6.2: Characters for three different vibrations under C_{2v} point group.

C_{2v}	E	C_2	σ_{xz}	σ_{yz}	
Γ_1	1	1	1	1	$= A_1$
Γ_2	1	–1	–1	1	$= B_2$
Γ_3	1	1	1	1	$= A_1$

Therefore, the final conclusion is that the symmetric stretching vibration of water molecule is of A_1 symmetry.

6.2.2 Antisymmetric Stretching Mode

In a similar way, the effect of different operations are carried out on the bond vectors corresponding to antisymmetric stretching and the nature of phase before and after symmetry operations are checked and characters are tabulated in Table 6.2 (see Γ_2) and pictorially it is presented in Figure 6.3.

It is seen that under few operations, the phase changes and phase remains unaltered under few operations. When the characters are tabulated corresponding to all the symmetry operations of C_{2v} point group, it is found that the characters of antisymmetric stretching mode of vibrations ultimately take the form according to B_2-symmetry (see character table).

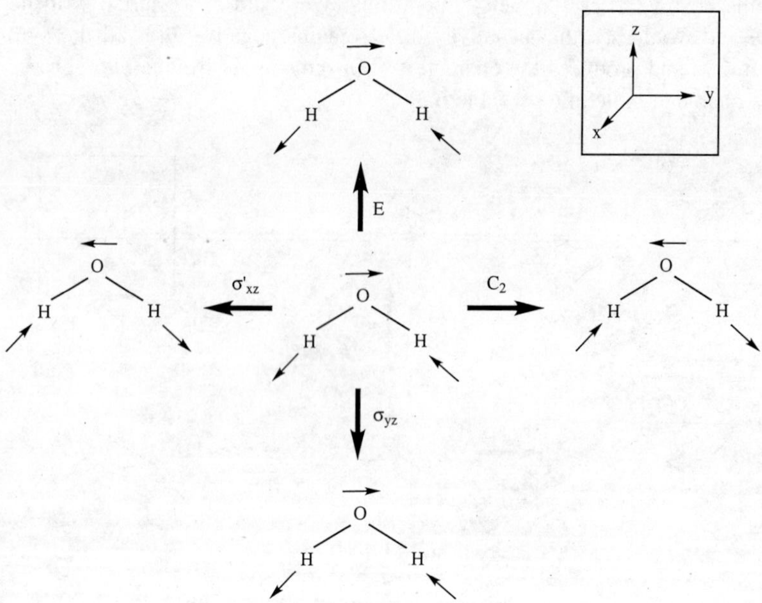

Figure 6.3: Effect of symmetry operations on anti-symmetric stretching mode of vibration.

6.2.3 Bending Mode

Similar exercise is done for the bending mode as well and shown in Figure 6.4 and the corresponding characters (Γ_3) are presented in Table 6.2.

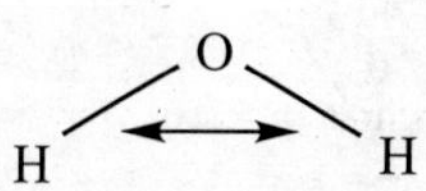

Bending mode of vibration

Alternatively, we can use a double headed arrow as shown above to construct the representation as well.

It is seen that under all the operations, the bending mode of vibration remains unchanged and transformed as A_1 symmetry (see character table).

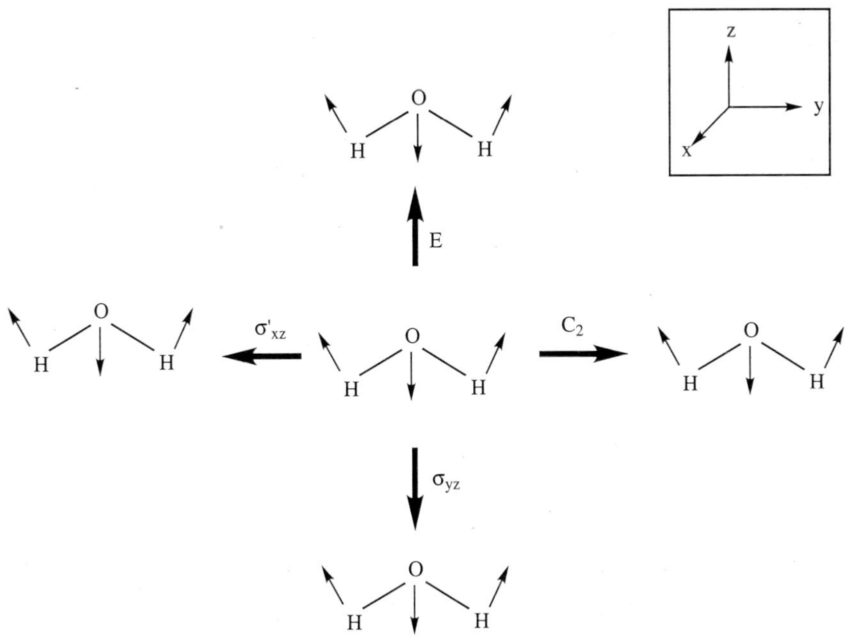

Figure 6.4: Effect of symmetry operations on bending mode of vibration.

6.3 NH_3 Molecule (C_{3v} Point Group)

Various symmetry elements of NH_3 molecule are E, $2C_3$, $3\sigma_v$. Cartesian coordinates for each atom is taken as basis to describe the total movements of the molecule. We have learnt how to get the reducible representation corresponding to a point group taking all the 12 coordinates for four atoms as basis. We can tabulate the reducible representation as shown in Table 6.3.

Table 6.3: Reducible representation of C_{3v} point group.

C_{3v}	E	$2C_3$	$3\sigma_v$
Γ_{Red}	12	0	2

Taking the help of character table, we can reduce the reducible representation into its corresponding irreducible components as:

$$3A_1 + A_2 + 4E$$

Now, from character table we can find out the components associated to translation and rotation movements $(A_1 + A_2 + 2E)$. Hence, we are left with combination of different irreducible representations corresponding to various vibrational movements as $2A_1 + 2E$. Therefore, we get total six vibrations with two totally symmetric, A_1 symmetry types and two doubly degenerate (total 4) of E symmetry types.

Using our common sense and chemistry knowledge, we can draw the different vibrations as shown in Figure 6.5. From the figures of symmetric stretching and symmetric bending, it is evident that under the six symmetry operations of C_{3v}, the

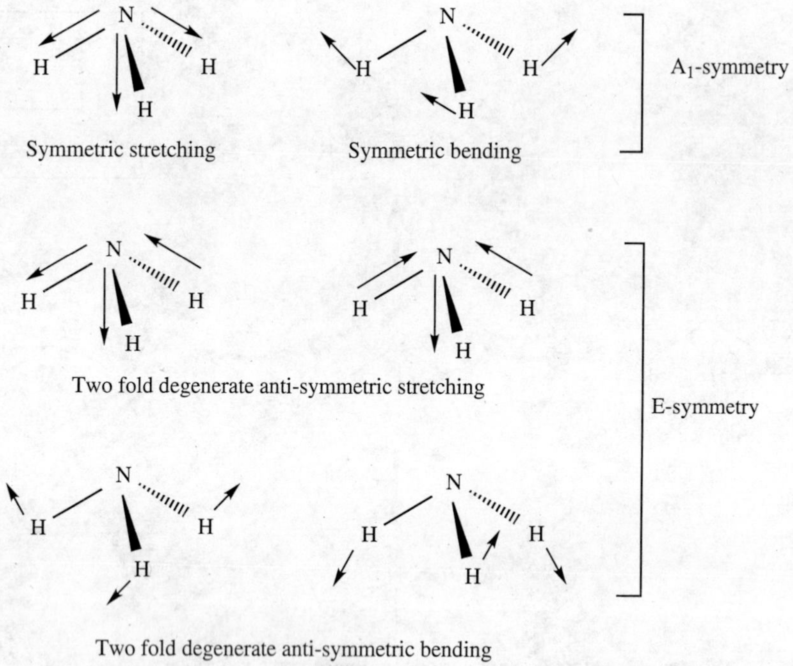

Figure 6.5: Movements of each atom during various type of vibrations of NH_3 molecule.

direction of arrows remain unaltered, which corresponds to A_1 symmetry. Other two sets of antisymmetric stretching and antisymmetric bending modes, with their atomic movements, have been shown in the figure as degenerate E-symmetry.

6.4 BCl_3 Molecule (D_{3h} Point Group)

In BCl_3 molecule, the various symmetry elements are E, $2C_3$, $3C_2$, σ_h, $2S_3$, and $3\sigma_v$. Taking Cartesian coordinates as basis, we get the following reducible representation shown in Table 6.4.

Table 6.4: Reducible representation of D_{3h} point group.

D_{3h}	E	$2C_3$	$3C_2$	σ_h	$2S_3$	$3\sigma_v$
Γ_{Red}	12	0	−2	4	−2	2

Using reduction formula, we get the constituent IRs as:

$$\Gamma_{Red} = A_1' + A_2' + 3E' + 2A_2'' + E''$$

Subtracting the irreducible representations due to translation and rotation movements (obtained from character table), we get the representation corresponding to only vibrational motion as:

$$\Gamma_{Vib} = A_1' + 2E' + A_2''$$

There are total 6 vibrational modes, out of which two are of nondegenerate A_1' (totally symmetric) and A_2'' symmetry and two doubly degenerate, E' symmetry.

6.5 CH_4 Molecule (T_d Point Group)

For five atoms, a set of 15 Cartesian displacement vectors form a basis to produce the reducible representation for this molecule. The list of symmetry operations along with the reducible representation is listed in Table 6.5.

Table 6.5: Reducible representation of T_d point group.

T_d	E	$8C_3(4C_3, 4C_3^2)$	$3C_2$	$6S_4(3S_4, 3S_4^3)$	$6\sigma_d$
Γ_{Red}	15	0	–1	–1	3

Using character table and applying reduction formula, this reducible representation is reduced to

$$\Gamma_{Red} = A_1 + E + T_1 + 3T_2$$

In a similar way, as earlier,from character table we find that translational motions transform as T_2 and rotations transform as T_1. Hence, the remaining symmetries are due to different vibrations, *i.e.*,

$$\Gamma_{vib} = A_1 + E + 2T_2$$

Therefore, here it is seen that there are total 9 mode of vibrations and out of which two are triply degenerate (That means total six vibrations are there which have the same frequency) of T_2 type symmetry, one doubly degenerate (Two vibrations of same frequency) with E-symmetry type and one vibration with totally symmetric type, A_1. From our basic common sense we can describe very easily this A_1 type of movement. Totally symmetric means all the atoms are moving in the same phase, *i.e.*, all the atoms are expanding or contracting simultaneously. Later we will learn to describe the detail and exact movement of each and every atom using projection operation method for other vibrations as well.

6.6 AB_6 Type of Molecule (O_h Point Group)

Let us consider one AB_6 molecule which is octahedral in geometry where six bond vectors a_1, a_2, a_3, a_4, a_5 and a_6 (according to the numbering sequence of peripheral atom, B)along each A-B bond are considered as basis and will try to find the reducible representation. Unlike the earlier examples, here each atom is connected to other atoms through five or six bonds. Hence, it is obvious that here the movement of the atoms are restricted to a large extent. Therefore, it would not be very wise decision to take three vectors for each atom along three cartesian coordinates when we are trying to get information about only different bond stretching movements of the molecule. At this point, just only counting the shifted and unshifted basis vectors against the different symmetry operations of this point group, the following table (Table 6.6) for reducible representation can be constructed.

Using character table and applying reduction formula, we get the number of times any IR appears in a reducible representation as,

$$a_i = \frac{1}{h}\left[\sum_{all\ classes} n_R.\chi(R).\chi_i(R)\right]$$

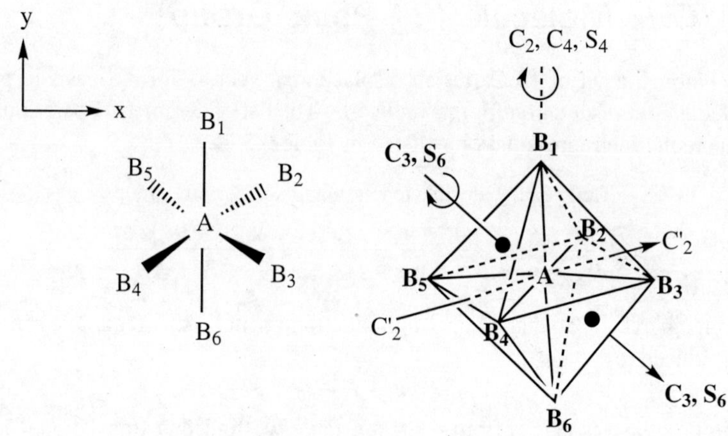

Figure 6.6: Octahedral AB_6 molecule with its different rotational axes.

Table 6.6: Reducible representation of O_h point group.

O_h	E	$8C_3$	$6C_2'$	$6C_4$	$3C_2$	i	$6S_4$	$8S_6$	$3\sigma_h$	$6\sigma_d$
$\Gamma_{A-B\,bond}$	6	0	0	2	2	0	0	0	4	2

Hence,

$$\text{Number of } A_{1g} : \frac{1}{48}[(1\times 6)+6(1\times 2)+3(1\times 2)+3(1\times 4)+6(1\times 2)]=1$$

$$\text{Number of } A_{2g} : \frac{1}{48}[(1\times 6)+6(-1\times 2)+3(1\times 2)+3(1\times 4)+6(-1\times 2)]=0$$

$$\text{Number of } E_{g} : \frac{1}{48}[(2\times 6)+6(0\times 2)+3(2\times 2)+3(2\times 4)+6(0\times 2)]=1$$

$$\text{Number of } T_{1g} : \frac{1}{48}[(3\times 6)+6(1\times 2)+3(-1\times 2)+3(-1\times 4)+6(-1\times 2)]=0$$

$$\text{Number of } T_{2g} : \frac{1}{48}[(3\times 6)+6(-1\times 2)+3(-1\times 2)+3(-1\times 4)+6(1\times 2)]=0$$

$$\text{Number of } A_{1u} : \frac{1}{48}[(1\times 6)+6(1\times 2)+3(1\times 2)+3(-1\times 4)+6(1\times 2)]=0$$

$$\text{Number of } A_{2u} : \frac{1}{48}[(1\times 6)+6(-1\times 2)+3(1\times 2)+3(-1\times 4)+6(1\times 2)]=0$$

$$\text{Number of } E_{u} : \frac{1}{48}[(2\times 6)+6(0\times 2)+3(2\times 2)+3(-2\times 4)+6(0\times 2)]=0$$

$$\text{Number of } T_{1u} : \frac{1}{48}[(3\times 6)+6(1\times 2)+3(-1\times 2)+3(1\times 4)+6(1\times 2)]=1$$

$$\text{Number of } T_{2u} : \frac{1}{48}[(3\times 6)+6(-1\times 2)+3(-1\times 2)+3(1\times 4)+6(-1\times 2)]=0$$

Hence, it is found that, $\Gamma_{A-B\,bond} = A_{1g} + E_g + T_{1u}$.

Here, it is evident that total six type of A-B stretching movements are possible, out of which one A-B bond stretching is of totally symmetric, A_{1g} symmetry, two are doubly degenerate of E_g symmetry and three are triply degenerate of T_{1u} symmetry.

Now, in other way, for the same AB_6 molecule, if we need to address any problem related to the movements of each atom in three dimensional directions (*e.g.*, vibrational movement), we can take the three Cartesian coordinates of each seven atom as basis and can find reducible representation corresponding to total degrees of freedom and then by subsequent reduction IR components are developed. In that case the reducible representation can be constructed as shown in Table 6.7:

Table 6.7: Reducible representation of O_h point group.

O_h	E	$8C_3$	$6C_2'$	$6C_4$	$3C_2$	i	$6S_4$	$8S_6$	$3\sigma_h$	$6\sigma_d$
Γ	21	0	−1	3	−3	−3	−1	0	5	3

And then Γ is reduced to,

$$\Gamma = A_{1g} + E_g + T_{1g} + 3T_{1u} + T_{2g} + T_{2u}$$

This describes all the possible type of movements of all the atoms. We have already learnt how to figure out the translational and rotational movements of a molecule from a knowledge of character table. If we subtract these two contributions, we can very easily find out the symmetry nature of all possible vibrational mode of this molecule. Then further advanced treatment (like projection operation technique) can give the detail and exact picture of a particular vibrational mode.

In this case, character table depicts that one T_{1g} and one T_{1u} corresponds to translational and rotational degrees of freedom respectively. Hence, all possible symmetry type of various vibrational modes corresponds to

$$\Gamma_{Vib} = A_{1g} + E_g + 2T_{1u} + T_{2g} + T_{2u}$$

Hence, there are total 15 various types of vibrational modes present for AB_6 type of molecule.

In this way, by choosing proper basis, the symmetry of different vibrational modes can easily be determined for any molecule.

EXERCISES

6.1 Determine the number of vibrational frequencies, their symmetries, and IR and Raman activities of the various vibrational modes of the following molecules: (i) NH_3 (ii) *staggered*-Ethane (iii) PF_5 (iv) Pyridine.

6.2 Determine the number of vibrational frequencies, their symmetries, and the IR and Raman activities of the various vibrational modes of following molecules. Comment on their polarization of the Raman bands as well: (i) CO_2 (ii) OCN^- (iii) Acetylene (iv) *trans*-N_2F_2 (v) SF_6.

6.3 Determine the nature of symmetry of the bending mode of vibrations for CO_2 molecule.

6.4 Which irreducible representations correspond to the stretching mode of vibrations of XeF_4 molecule?

6.5 Show that in CH_4 molecule, two such vibrations are there which are triply degenerate.

Chapter 7

Application of Symmetry for Determination of IR and Raman Activity and Assignments of Spectral Data

7.1 Introduction

It is known that a molecule to be Infrared (IR) active, there must be a change in dipole moment during vibration. Details of this kind of spectroscopy is available in any spectroscopy book. At this point, we can have a very brief discussion of Raman spectroscopy as well. Unlike IR spectroscopy, Raman spectroscopy is an indirect technique which involves irradiation of sample with high intensity radiation from a laser source and measuring the interaction of the sample with the radiation, at right angles to the input direction. Normally, photons of radiation will only excite a molecule if there is a matching in energy between a pair of energy levels, like IR spectroscopy. If such an energy matching does not occur, the photon and the sample interact elastically, and no energy is transferred. However, in a very small number of cases (1 in 10^7) the interaction is inelastic, and energy is transferred between the photon and the sample. To measure these rare events, the intensity of the incident radiation must be high and this criterion is fulfilled by the use of laser. Moreover, it is difficult to measure these rare events against a background of the overwhelming majority of elastic collisions. Hence, the scattered radiation is measured at right angles to the direction of incident radiation *i.e.*, against a blank background.

The amount of energy transferred ($\Delta E = h\nu$), is only a fraction of the total energy of the photon ($\Delta E - h\nu'$) and corresponds to excitation of a particular vibrational mode. The energy of the scattered photon gets reduced and occurs at a new frequency, ν'' when the radiation is absorbed by molecule. Another possibility, though less common, is the interaction of photon with a molecule existing already in an excited vibrational state. Here, energy is lost by the molecule during its return to vibrational ground state and consequently the frequency (energy) of the scattered photon increases to $\nu' + \nu''$. Raman spectrum measures the changes in the frequency of radiation, $\nu' + \nu''$ or $\nu' - \nu''$, which corresponds to the energy gap between vibrational states, ν. Unlike IR spectroscopy, here the incident radiation is of a single frequency (monochromatic) and the vibrational spectrum is measured indirectly.

The selection rule for Raman spectroscopy is that:

A vibrational mode is only Raman active if it results in a change in molecular polarisibility.

Polarisibility describes the ease or difficulty with which an electron cloud can be distorted and has the dimension of volume. In the case of the symmetric stretching of CO_2, the polarisibility changes, as both the bonds simultaneously either shorten or lengthen and hence volume changes as well. Thus, while the symmetric stretching mode of CO_2 is IR inactive, is Raman active.

So far we have determined the symmetries of different possible vibrations of a molecule. Now we have to find out which vibrations are IR/Raman active and which are inactive among all those vibrations feasible for a molecule, in terms of symmetry property.

For this we have two simple rules to obey.

1. *A vibration will be IR active if it belongs to the same symmetry species as a component of dipole moment i.e., to the same symmetry species as either x,y or z (see character tables).*
2. *A vibration will be Raman active if it belongs to the same symmetry species as a component of polarisibility i.e., to one of the binary products, $x^2, y^2, z^2, xy, yz, xz$ or to a combination of products e.g. $x^2 - y^2$. (see character tables).*

Let us take few examples to discuss the topic.

7.1.1 H_2O Molecule (C_{2v})

Already we have learnt how to determine the symmetry of different vibrations of a molecule. For water molecule, we have seen that there are three vibrations and among them, two are of totally symmetric, A_1 type and rest one is of B_2 symmetry type, *i.e.*,

$$\Gamma_{\text{Vib}} = 2A_1 + B_2$$

Recalling the above rules and character table (Table 7.1) of C_{2v} point group, it is found that all the three vibrations are IR active, since A_1 and B_2 irreducible representations contain either y or z as basis. More careful inspection shows that z belongs to A_1 and y belongs to B_2 symmetry. This implies that the change in dipole moment of A_1 symmetry type vibrations (symmetric stretching and bending as already seen in our previous discussion) occur along z direction, also called *z-polarized vibration* and dipole moment of B_2 type of vibration (asymmetric stretching) changes along y direction, also called *y-polarized vibration*.

Table 7.1: Character table for C_{2v} point group and corresponding symmetry of functions.

C_{2v}	E	C_2	$\sigma(xz)$	$\sigma'(yz)$	Symmetry of functions	
A_1	1	1	1	1	z	x^2, y^2, z^2
A_2	1	1	–1	–1	R_z	xy
B_1	1	–1	1	–1	x, R_y	xz
B_2	1	–1	–1	1	y, R_x	yz

The meaning of polarization can be nicely illustrated by taking experimental IR data of water molecule as an example. The three IR frequencies of water molecule are $\nu_1(3657\text{cm}^{-1})$, $\nu_2(1595\ \text{cm}^{-1})$ and $\nu_3(3756\ \text{cm}^{-1})$. According to convention, ν_1, ν_2 are named after symmetric stretching and bending respectively, of a_1 symmetry, and ν_3 is designated as antisymmetric stretching of b_2 symmetry. From transition dipole moment calculations, this can be recalled that ν_1 and ν_2 mode of vibrations are z-polarized and ν_3 is y-polarized.

During a hypothetical measurement of IR spectra of water molecule, if depolarized light is used (where all the three x, y and z electric components are present), all the three vibrational frequencies are observed; in presence of z-polarized light, only z-polarized vibrations are observed; in presence of y-polarized light, only signature of y-polarized, asymmetric vibration is observed and as expected in presence of x-polarized light, no spectral signature is observed since, none of the vibrations are x-polarized.

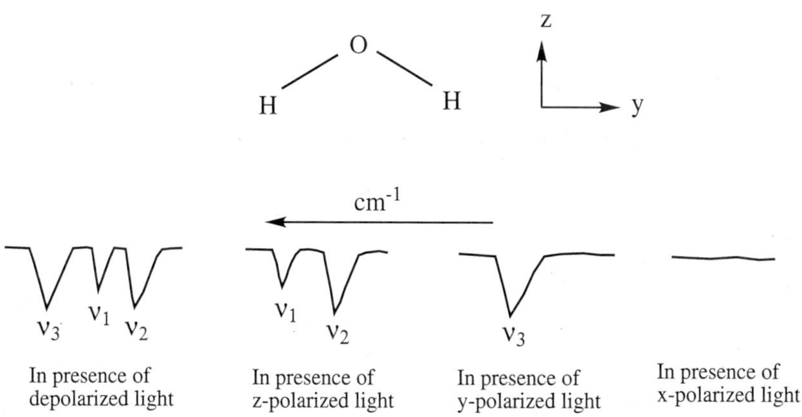

Figure 7.1: Hypothetical IR spectra of water molecule in presence of different polarized light.

Similarly, all the three vibrations are Raman active since, x^2, y^2 and z^2 binary components belong to A_1 and xz component belongs to B_2 symmetry.

7.1.2 NH_3 Molecule (C_{3v})

Taking the help of character table, we can generate the reducible representation (RR) taking all possible movements of the atoms of the molecule as basis and reduce this representation into its corresponding components as:

$$\Gamma_{Total} = 3A_1 + A_2 + 4E$$

Subtracting translation and rotation contributions we get,

$$\Gamma_{Vib} = 2A_1 + 2E$$

The character table of C_{3v} is recalled below in Table 7.2 for our ready reference.

Table 7.2: Character table for C_{3v} Symmetry.

C_{3v}	E	$2C_3$	$3\sigma_v$	*Symmetry of functions*
A_1	1	1	1	z, x^2+y^2, z^2
A_2	1	1	−1	R_z
E	2	−1	0	$(x,y)(R_x,R_y),(x^2+y^2,xy)(yz,xz)$

Already we have seen that for NH_3 molecule symmetric stretching and symmetric bending vibrations are of A_1 type and since this irreducible representation contains component of x, y or z (here z) as well as binary components (here x^2+y^2 and z^2), hence these two type of vibrations are Infrared as well as Raman active.

Similarly, since corresponding to E irreducible representation there are x,y as well as binary components, hence two doubly degenerate vibrations (two antisymmetric stretching and two antisymmetric bending) are IR as well as Raman active.

7.1.3 CH_4 Molecule (T_d)

Taking the help of character table, we can get the reducible representation (Γ_{Total}), taking all possible movements of the atoms of the molecule, as basis and reduce this representation into its corresponding irreducible components as (also see section 6.5):

$$\Gamma_{Total} = A_1 + E + T_1 + 3T_2$$

Subtracting translation and rotation contributions, we get,

$$\Gamma_{vib} = A_1 + E + 2T_2$$

Now, out of these total 12 different mode vibrations, from character table, if we follow the rules, we can conclude that vibrations corresponding to A_1 and E symmetry are IR inactive, but 6 vibrations (two triply degenerate) corresponding to T_2 symmetry are IR active. Also we can conclude that except T_2 symmetry vibrations, in other type of vibrations (A_1 and E) there is no change in dipole moment.

Also it can be concluded that vibrations corresponding to all the three different symmetries are Raman active as in all irreducible representations, there are various binary functions (see character table). This indicates that during all the different vibrations there occur change in polarisibility.

Before proceeding further to assign spectral data, let us have a very brief discussion about the *Polarized* and *Depolarized* nature of Raman active vibrational mode.

7.2 Polarized and Depolarized Nature of Raman Active Vibrational Mode

Electromagnetic wave can have any orientation in space depending upon its nature. If, in a Raman experiment, the incident radiation is confined to only one plane (*i.e., plane polarized*) by the use of a slit, rather than randomly oriented, then the interactions with the molecule, under study, will vary from one molecular vibration to another. Let us consider the symmetrical stretching of a spherically symmetrical molecule, such as CCl_4. It is seen that the polarizibility ellipsoid (the measure of change of polarisibility in three dimensional space) of the molecule will change, since the bonds stretch or contract simultaneously. However, as this expansion or contraction is perfectly symmetrical, the polarizibility, though different, remains same in all directions, *i.e.*, although the size of polarisibilty ellipsoid changes but shape remains same. It does not matter what orientation the CCl_4 molecule adopts in space, the polarisibility of the molecule, as seen by the incident plane-polarised radiation, will always remain the same. The net effect is that, in an interaction with a species undergoing a perfectly symmetrical vibration, plane-polarised radiation remains polarized after scattering by the sample as shown in Figure 7.2a below.

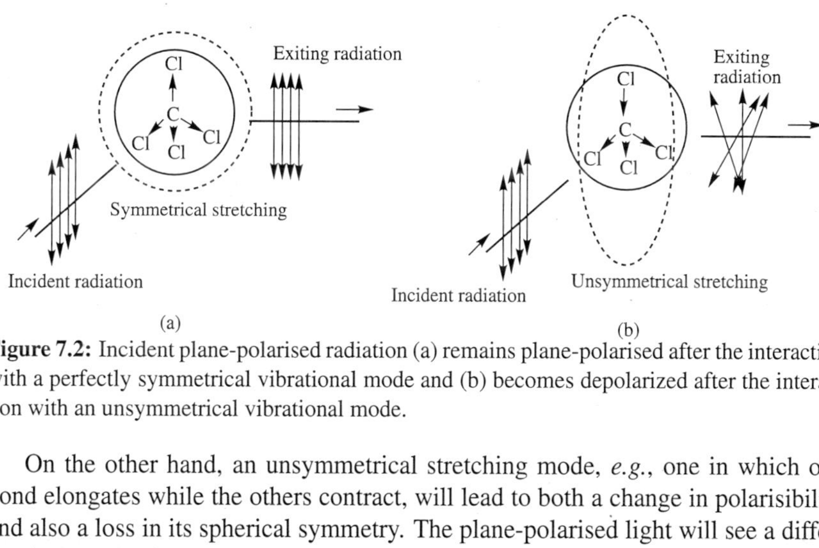

Figure 7.2: Incident plane-polarised radiation (a) remains plane-polarised after the interaction with a perfectly symmetrical vibrational mode and (b) becomes depolarized after the interaction with an unsymmetrical vibrational mode.

On the other hand, an unsymmetrical stretching mode, *e.g.*, one in which one bond elongates while the others contract, will lead to both a change in polarisibility and also a loss in its spherical symmetry. The plane-polarised light will see a different degree of polarisibilty depending on how CCl_4 molecule is oriented towards the incident plane-polarised radiation. The result of this is that, despite starting off as plane-polarised, due to the presence of all possible orientations of the exiting radiation, the wave becomes depolarized after scattering by an unsymmetrical vibrational mode (Figure 7.2b).

This leads to the following general observations:

1. *A Raman mode will only be polarized if it is perfectly symmetrical.*
2. *A Raman polarized mode corresponds to +1 in the character table for all symmetry operations—the totally symmetric representation, and always corresponds to the first row of the irreducible representations listed.*

7.2.1 Correlation between Nature of Vibrational Mode and Nature of Raman Spectra

Whether a given Raman signal is arising from a totally symmetric vibration or not can be determined by observing the polarisation of the exiting signal. The experimental procedure is to use a plane-polarised light beam (say $I_{\|}$) as incident (exciting) radiation and collect the scattered Raman signal intensities in two directions, namely $I_{\|}$ (Parallel) and $I_{\perp}$ (perpendicular) simultaneously with respect to the incident beam. The intensity ratio, thus obtained, $\rho = \frac{I_{\perp}}{I_{\|}}$, is called the *depolarisation ratio*. If ρ is close to zero (*i.e.*, $I_{\perp} \sim 0$), means that the Raman scattered light is in same state of polarisation as that of the incident light and the vibrational mode responsible for this Raman signal is totally symmetric in nature. The totally symmetric mode is designated depending upon the nature of point group of a molecule and is identified either by A or A_1 or A_g from character table.

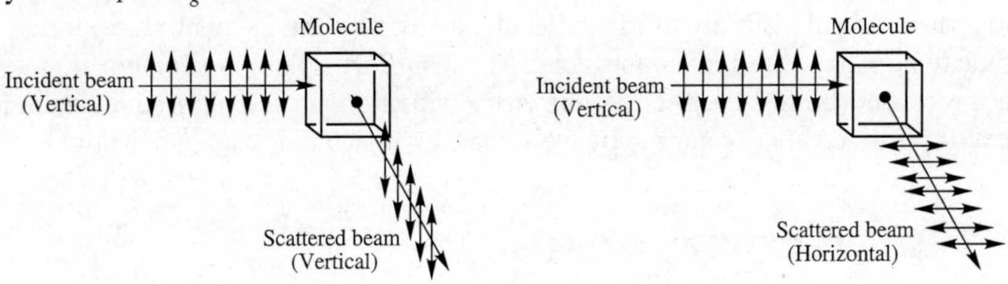

Parallel polarization ($I_{\|}$)
(a)

Perpendicular polarization ($I_{\perp}$)
(b)

Figure 7.3: Illustration of Parallel and Perpendicular polarization.

Higher values of ρ, indicates the presence of other state of polarization and hence, scattered light is to be *depolarised* and the vibration is considered as *totally symmetric*. It is found that if $0 \leq \rho \leq 0.75$, the line is taken as *polarised* (*i.e.*, totally symmetric vibration). For $\rho > 0.75$, the line is taken as depolarised and hence the corresponding vibration is considered as *non-totally symmetric* in nature.

If a plane polarised light beam is incident on a spherical top, the plane of the induced dipole in the case of a totally symmetric vibration is always parallel to that of the incident light and the molecule behaves like tiny sphere. The scattered light is also acquire polarisation parallel to the plan of the incident light, making $\rho \sim 0$. Molecules with T_d or O_h geometry behave in the same way. For T_d molecule, out of the five symmetries (A_1, A_2, E, T_1 and T_2), A_1, E and T_2 are Raman active and only T_2 is IR active as well and from experimental data for A_1 band $\rho \sim 0$, whereas for E and T_2 bands $\rho \sim 1$.

7.3 Application of Symmetry for Assignments of Spectral Data

For detail assignment of symmetry for spectral bands, we need to remember few simple general rules:

- *In a given molecule, stretching frequencies are higher than bending frequencies.*
- *Among the stretching modes of vibration, asymmetric stretchings are of higher frequency than symmetric ones.*
- *For molecules of symmetric top type, the Raman transitions corresponding to the totally symmetric irreducible representation (which occurs at the top of the list of irreducible representation in character table, like A, A_1, A_g etc.) must be polarised.*
- *Out-of-plane bendings are generally of higher frequencies than in-plane bendings.*
- *For bipyramidal type of molecule where two different type of bonds are there (axial and equatorial), the bending vibration involving two equatorial bonds is found to be at lower frequency than bending involving one axial and one equatorial bond.*

Let us take few examples to illustrate these rules:

7.3.1 *Trans*-N_2F_2 (C_{2h} Point Group)

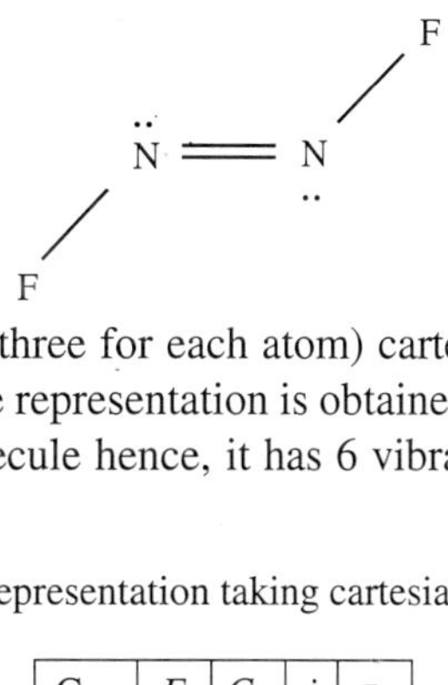

If we take the set of 12 (three for each atom) cartesian displacement vectors as basis, the following reducible representation is obtained, presented in Table 7.3. It is a four atomic nonlinear molecule hence, it has 6 vibrational degrees of freedom or internal freedom.

Table 7.3: Reducible representation taking cartesian displacement vectors.

C_{2h}	E	C_2	i	σ_h
Γ_{red}	12	0	0	4

This reducible representation can be reduced as:

$$\text{Number of } A_g : \frac{1}{4}[(1\times 12)+(1\times 0)+(1\times 0)+(1\times 4)] = 4$$

$$\text{Number of } B_g : \frac{1}{4}[(1\times 12)+(-1\times 0)+(1\times 0)+(-1\times 4)] = 2$$

$$\text{Number of } A_u : \frac{1}{4}[(1\times 12)+(1\times 0)+(-1\times 0)+(-1\times 4)] = 2$$

$$\text{Number of } B_u : \frac{1}{4}[(1\times 12)+(-1\times 0)+(-1\times 0)+(1\times 4)] = 4$$

Therefore, the reducible representation can be expressed as,

$$\Gamma_{red} = 4A_g + 2B_g + 2A_u + 4B_u$$

Inspection of character table shows that total contribution due to translational and rotational motion is $A_g + 2B_g + A_u + 2B_u$.

Table 7.4: C_{2h} point group along with associated functions.

C_{2h}	E	C_2	i	σ_h	*Functions*	
A_g	1	1	1	1	R_z	$(x^2, y^2, z^2), xy$
B_g	1	−1	1	−1	R_x, R_y	xz, yz
A_u	1	1	−1	−1	z	
B_u	1	−1	−1	1	x, y	

By removing translational and rotational contributions, symmetry of total six vibrations are obtained,

$$\Gamma_{vib} = 3A_g + A_u + 2B_u$$

Further, taking the help from character table (Table 7.4), it is obvious that vibrations corresponding to A_u and B_u symmetries are IR active, whereas, only A_g symmetry vibrations are Raman active.

Taking various internal coordinates, the nature of those 6 vibrations can be found. At this point, it may be noted that A_g and B_u vibrations must involve *in plane* movements, as from character table, we can see that those two kind of vibrations remain unchanged with respect to σ_h operation (+1 character). But, A_u kind of vibration involves *out-of-plane* deformation since, it shows –1 character under σ_h operation. Hence, we can conclude that out of the six vibrations **at least one must be out-of-plane** and **five are of in-plane vibrations**. To get more precise conclusions about the movements of those five vibrations, we have to choose five internal coordinates which can describe only planar movements. The set is comprised of two N−F bonds, two $\angle NNF$ angles and one N=N bond distance. Each of these basis produces corresponding reducible representations as:

Table 7.5: Reducible representations using different bases.

C_{2h}	E	C_2	i	σ_h
Γ_{NF}	2	0	0	2
Γ_{NNF}	2	0	0	2
Γ_{NN}	1	1	1	1
$\Gamma_{\text{out-of-plane}}$	2	0	0	−2

After reduction, for NF bond vector as basis it is found as:

$$\text{Number of } A_g : \frac{1}{4}[(1\times 2)+(1\times 0)+(1\times 0)+(1\times 2)] = 1$$

$$\text{Number of } B_g : \frac{1}{4}[(1\times 2)+(-1\times 0)+(1\times 0)+(-1\times 2)] = 0$$

$$\text{Number of } A_u : \frac{1}{4}[(1\times 2)+(1\times 0)+(-1\times 0)+(-1\times 2)] = 0$$

$$\text{Number of } B_u : \frac{1}{4}[(1\times 2)+(-1\times 0)+(-1\times 0)+(1\times 2)] = 1$$

Hence,

$$\Gamma_{N-F} = A_g + B_u$$

Similarly,

$$\Gamma_{\angle NNF} = A_g + B_u, \quad \Gamma_{N-N} = A_g \text{ and } \Gamma_{\text{out-of-plane}} = A_u + B_g$$

Since, in the list of symmetries of total vibrationas, there was no vibration of B_g symmetry hence, B_g is considered to be redundant and it corresponds to R_x and R_y functions.

Therefore, it can be concluded that the three Raman active modes of vibrations (A_g) are composed of symmetric N−F stretching, symmetric $\angle NNF$ bending and N=N stretching. At this point it should be noted that the relative contribution of each movement in each normal mode depends on the force constants and atomic masses of participating atoms.

Similarly, the two B_u symmetry vibrations in IR spectroscopy are consisting of asymmetric N-F stretching and asymmetric $\angle NNF$ bending and again the proportion of these two movements in each normal mode will depend upon the force constant and atomic masses. From the available experimental data and group theoretical treatment for this molecule, the following assignments can be done.

Table 7.6: Assignments of IR and Raman spectral data with comments.

IR (gas, cm^{-1})	**Raman (gas, cm^{-1})**	**Assignment**	**Comments**
	1636 (pol)	A_g : N=N Stretching	* IR and Raman are mutually exclusive * A_g is Raman active and polarized. * Energy order: N=N str > N-F str > Bending
	1010 (pol)	A_g : N-F Stretching	
989		B_u : N-F Stretching	
	592 (pol)	A_g : N-N-F Deformation	
412		A_u or B_u: Out-of-plane Deformation/N-N-F angle deformation	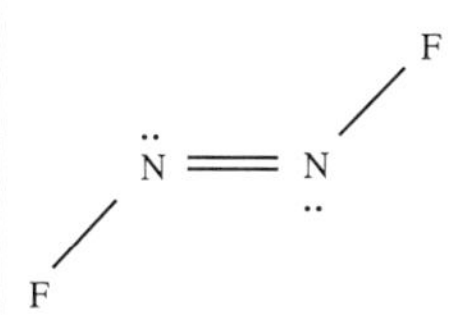
360		A_u or B_u: Out-of-plane Deformation/N-N-F angle deformation	

7.3.2 CH_4 (T_d Point Group)

For five atoms a set of 15 cartesian displacement vectors form a basis to produce the reducible representation shown in Table 7.7.
This is further reduced to:

$$\Gamma = A_1 + E + T_1 + 3T_2$$

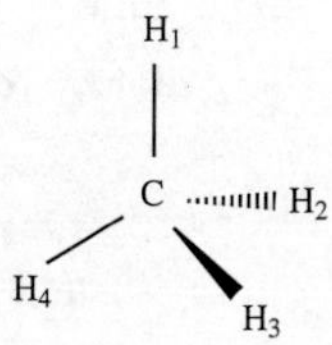

Table 7.7: Reducible representation of T_d molecule.

T_d	E	$8C_3(4C_3, 4C_3^2)$	$3C_2$	$6S_4\ (3S_4, 3S_4^3)$	$6\sigma_d$
Γ	15	0	–1	–1	3

From character table, we find that translation transforms as T_2 and rotation transforms as T_1. Hence, the remaining symmetries are due to vibration or internal freedom as:

$$\Gamma_{vib} = A_1 + E + 2T_2$$

(Total 9 vibrations; two triply degenerate, one doubly degenerate and one A_1 type).

Table 7.8: Character table for the T_d point group.

T_d	E	$8C_3$	$3C_2$	$6S_4$	$6\sigma_d$	*Symmetry of functions*	
A_1	1	1	1	1	1		$x^2+y^2+z^2$
A_2	1	1	–1	–1	–1		
E	2	–1	2	0	0		$(2z^2-x^2-y^2, x^2-y^2)$
T_1	3	0	–1	1	–1	(R_x, R_y, R_z)	
T_2	3	0	–1	–1	1	(x, y, z)	(xy, yz, xz)

The above character table shows that A_1 and E symmetry motions are Raman active and T_2 symmetry corresponds to both IR and Raman activity.

To find the contributions of the internal coordinates, four C–H bonds are taken as basis and reducible representation is generated as shown in Table 7.9.

Table 7.9: Reducible representation using C–H bond vectors as basis.

$\mathbf{T_d}$	**E**	$\mathbf{8C_3}$	$\mathbf{3C_2}$	$\mathbf{6S_4}$	$\mathbf{6\sigma_d}$
Γ_{C-H}	4	1	0	0	2

After reduction it is found as

$$\Gamma_{C-H} = A_1 + T_2, \quad \text{(total 4 modes of vibrations)}$$

Similar way, six ∠HCH angles (H_1CH_2, H_1CH_3, H_1CH_4, H_2CH_3, H_3CH_4, H_2CH_4) are taken as basis and generated the reducible representation as shown in Table 7.10.

After reduction,

$$\Gamma_{\angle HCH} = A_1 + E + T_2, \quad \text{(Total 6 modes of vibrations)}$$

Table 7.10: Reducible representation taking $\angle HCH$ angles as basis.

T_d	E	$8C_3$	$3C_2$	$6S_4$	$6\sigma_d$
$\Gamma_{\angle HCH}$	6	0	2	0	2

At this point, one interesting observation may be noted. Instead of total **nine** internal freedoms or normal mode of vibrations, here we get total **ten** vibrational symmetries. In case of C$-$H stretching, we can see that there are four vibrational modes (one singly, A_1 and one triply degenerate, T_2) of A_1 and T_2 symmetry. A_1 corresponds to symmetric stretching when all the four bonds expand and contract simultaneously and hence this type of vibration is IR inactive but as volume changes during this type of vibration, polarisibilily changes and hence Raman active (already evident from character table). Similar analysis helps to conclude that triply degenerate T_2 mode corresponding to C$-$H stretching motions, is basically asymmetric stretching where simultaneously one C$-$H bond contracts and other three bonds expand and vice versa.

For $\angle HCH$ bending, we can find total 6 vibrations (one singly, A_1, one doubly, E and one triply degenerate, T_2). Inspecting very carefully, we see that the presence of A_1 symmetry during $\angle HCH$ bending is redundant, as all the six bond angles can not vary (increase or decrease) in the same way at the same time simultaneously (as A_1 mode is totally symmetric). If any five bond angles changes (say increases), then automatically the value of the 6th angle will be predetermined to a particular value (depending upon the values of the five angles) *i.e.*, cannot change arbitrarily. Therefore, A_1 type of symmetry is not possible during bond angle movements. Hence, it can be concluded that this type of vibration corresponds to purely C$-$H stretching, there will be no contribution from $\angle HCH$ bond angle deformation.

E type of vibration corresponds to purely bond angle deformation and is Raman active only. T_2 type of vibrations are both IR and Raman active and also have the contribution from bond stretching as well as bond angle deformations and their relative contribution on each normal mode will depend on the force constant and atomic masses.

7.3.3 SF_6 (O_h Point Group)

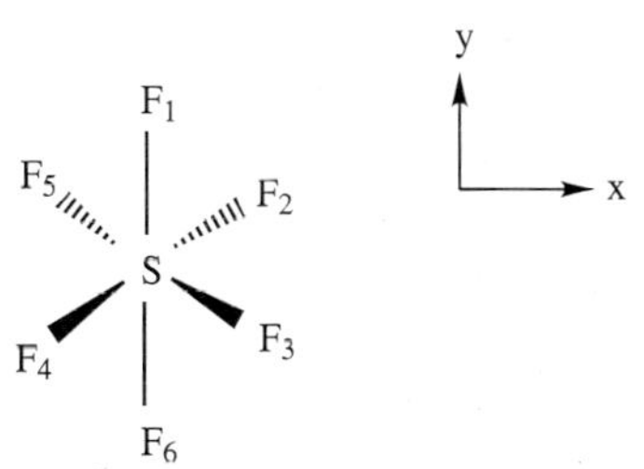

This molecule has 15 degrees of internal freedom or vibrational modes. The 21 cartesian displacement vectors generate the following representation shown in table below (Table 7.11):

Table 7.11: Reducible representation taking cartesian displacement vectors.

O_h	E	$8C_3$	$6C_2$	$6S_4$	$6C_2$	i	$6S_4$	$8S_6$	$3\sigma_h$	$6\sigma_d$
Γ_{Total}	21	0	–1	3	–3	–3	–2	0	5	3

Γ reduces as, $\Gamma_{\text{Total}} = A_{1g} + E_g + T_{1g} + 3T_{1u} + T_{2g} + T_{2u}$

At this point, taking the help of character table, if translation and rotation symmetries are subtracted (T_{1g} and T_{1u}), the remaining vibrational normal mode symmetries are as follows:

$$\Gamma_{\text{Vib}} = A_{1g} + E_g + 2T_{1u} + T_{2g} + T_{2u}$$

Again, from character table of O_h point group, shown below, we can group them according to their IR and Raman activities as:

IR active: $2T_{1u}$

Raman active: A_{1g}, E_g and T_{2g}

Inactive: T_{2u}

The occurrence of inactivity in both IR or Raman spectroscopies is not very common, but is found sometimes when the molecule possesses more symmetry.

Now, to go into more detail, we can take six S-F bonds and 12 $\angle FSF$ angles as basis for different vibrations to generate the reducible representation as shown in Table 7.12.

Table 7.12: Reducible representation taking S-F bonds and $\angle FSF$ angles as basis.

O_h	E	$8C_3$	$6C_2$	$6S_4$	$6C_2$	i	$6S_4$	$8S_6$	$3\sigma_h$	$6\sigma_d$
Γ_{SF}	6	0	0	2	2	0	0	0	4	2
Γ_{FSF}	12	0	2	0	0	0	0	0	4	2

After reduction we get,

$$\Gamma_{SF} = A_{1g} + E_g + T_{1u} \quad \text{and} \quad \Gamma_{\angle FSF} = A_{1g} + E_g + T_{1u} + T_{2g} + T_{2u}$$

Hence, we get 18 modes of vibrations instead of predicted correct number of 15. Since, bond stretching is independent to each other, hence, there can not be any redundancy in the representation, but since, on the contrary, all the bond angles can not change independently, there must be some redundancy as we have seen in case of T_d geometry. After careful inspection of O_h character table (Table 7.13), A_{1g} and E_g are found to be the two spurious symmetries. Hence, actual representation for $\Gamma_{\angle FSF}$ would be a combination of T_{1u}, T_{2g} and T_{2u}.

Therefore, it can be concluded that A_{1g} and E_g modes will involve only bond stretching, each of the two T_{1u} modes will involve a combination of bond stretching and bond angle deformation and the remaining two modes (T_{2g} and T_{2u}) will be comprised of only angle deformation.

Table 7.13: Character table of O_h point group.

O_h	E	$8C_3$	$6C_2$	$6S_4$	$6C_2$	i	$6S_4$	$8S_6$	$3\sigma_h$	$6\sigma_d$	Functions	
A_{1g}	1	1	1	1	1	1	1	1	1	1		$x^2+y^2+z^2$
A_{2g}	1	1	–1	–1	1	1	–1	1	1	–1		
E_g	2	–1	0	0	2	2	0	–1	2	0		$(2z^2-x^2-y^2, x^2-y^2)$
T_{1g}	3	0	–1	1	–1	3	1	0	–1	–1	R_x, R_y, R_z	
T_{2g}	3	0	1	–1	–1	3	–1	0	–1	1		xz, yz, xz
A_{1u}	1	1	1	1	1	–1	–1	–1	–1	–1		
A_{2u}	1	1	–1	–1	1	–1	1	–1	–1	1		
E_u	2	–1	0	0	2	–2	0	1	–2	0		
T_{1u}	3	0	–1	1	–1	–3	–1	0	1	1	(x, y, z)	
T_{2u}	3	0	1	–1	–1	–3	1	0	1	–1		

7.3.4 $Xe(O)F_4$ (C_{4v} Point Group)

VSEPR theory predicts an octahedral arrangement of six electron pairs with one site occupied by a lone pair of electron. The actual shape to be a square-based pyramid. Few experimental IR and Raman spectroscopy data are presented in Table 7.15.

Using three cartesian vectors for each atom, the reducible representation can be generated as shown in Table 7.14.

Table 7.14: Reducible representation using cartesian vectors as basis.

C_{4v}	E	$2C_4$	C_2	$2\sigma_v$	$2\sigma_d$
Γ	18	2	–2	4	12

After reduction we get,

$$\Gamma = 4A_1 + A_2 + 2B_1 + B_2 + 5E$$

Then after subtracting the rotation and translation (A_1, A_2, $2E$), we get,

$$\Gamma_{\text{vib}} = 3A_1 + 2B_1 + B_2 + 3E \quad \text{(Total 12 modes)}$$

From character table the following conclusion can be done regarding IR and Raman activities:

A_1 : IR active (T_z), Raman active (x^2+y^2, z^2, pol)
B_1 : Raman active (x^2-y^2, depol)
B_2 : Raman active (xy, depol)
E : IR active (T_x, T_y), Raman active (xz, yz, depol)

At this point, based on the information gathered, some initial assignments can be done which are presented in Table 7.16.

Table 7.15: Experimental IR and Raman data for $Xe(O)F_4$ molecule.

IR (vap, cm^{-1})	**Raman(liquid, cm^{-1})**
926	920 (pol)
609	605(depol)
576	567 (pol)
	527 (depol)
361	364 (depol)
288	286 (pol)
	232 (depol)
	220 (depol)
159	160 (depol)

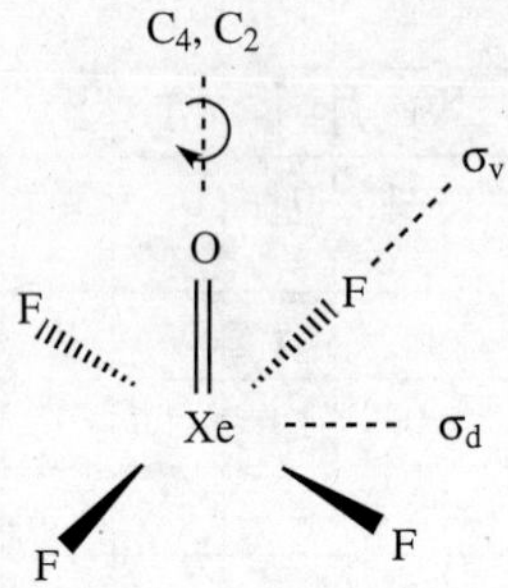

Table 7.16: Initial assignment of spectral bands.

IR (vap, cm^{-1})	**Raman(liquid, cm^{-1})**	**Assignment**	**Comments**
926	920 (pol)	A_1	* The A_1 and E modes are active in both IR and Raman. * only totally symmetric A_1 is Raman polarized. * B_1/B_2 are Raman –only mode but not IR active. *Raman mode will only be polarized if it is perfectly symmetrical during vibration.
609	605(depol)	E	
576	567 (pol)	A_1	
	527 (depol)	B_1 or B_2	
361	364 (depol)	E	
288	286 (pol)	A_1	
	232 (depol)	B_1 or B_2	
	220 (depol)	B_1 or B_2	
159	160 (depol)	E	

Now, to go into more details, let us construct the reducible representation for different bond stretching modes of either Xe-F or Xe-O bonds, shown in Table 7.17.

Table 7.17: Reducible representations for different bond stretching modes.

C_{4v}	E	$2C_4$	C_2	$2\sigma_v$	$2\sigma_d$
Γ_{Xe-O}	1	1	1	1	1
Γ_{Xe-F}	4	0	0	2	0

We can also consider the five bond vectors together as basis to take into account all the stretching modes. In that case, the reducible representation becomes as shown in Table 7.18.

Table 7.18: Reducible representation of C_{4v} point group.

C_{4v}	E	$2C_4$	C_2	$2\sigma_v$	$2\sigma_d$
Γ_{str}	5	1	1	3	1

On reduction of these stretching modes to its component IRs, we get,

$$\Gamma = 2A_1 + B_1 + E$$

Similarly, we can take the four $\angle FXeF$ angles and four $\angle FXeO$ angles separately as basis (represented by curved double headed arrows in Figure 7.4) and found the reducible representation as shown in Table 7.19.

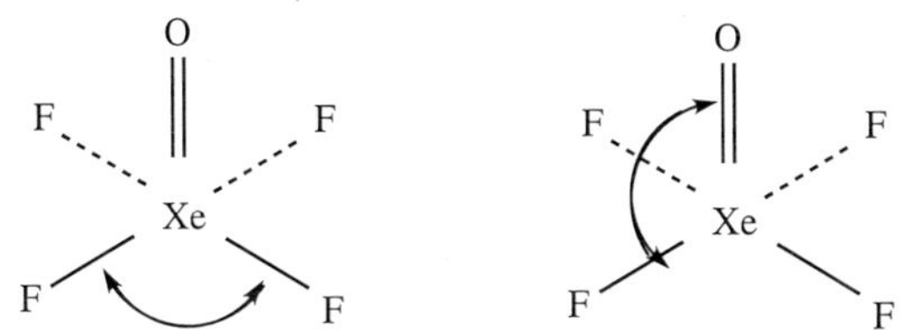

Figure 7.4: Equatorial and axial bond angles of $Xe(O)F_4$ molecule.

Table 7.19: Reducible representations, along with their reduced forms, taking different bond angles as basis.

C_{4v}	E	$2C_4$	C_2	$2\sigma_v$	$2\sigma_d$	After reduction
$\Gamma_{\angle FXeF}$	4	0	0	0	2	$= A_1 + B_2 + E$
$\Gamma_{\angle FXeO}$	4	0	0	2	0	$= A_1 + B_1 + E$

[Again, similar way, we can take all the eight angles together also and in that case the reducible representation will be:

Table 7.20: Reducible representation taking all the bond angles together as basis.

C_{4v}	E	$2C_4$	C_2	$2\sigma_v$	$2\sigma_d$
Γ_{angles}	8	0	0	2	2

And after reduction we will get,

$$\Gamma_{angles} = 2A_1 + B_1 + B_2 + 2E]$$

Therefore, it is found that instead of total 12 vibrational normal modes, from symmetry aspects, total 13 modes of vibrations (5 stretching and 8 angle deformations) are obtained. This indicates that there one extra mode of vibration (specifically A_1 mode) is coming out and that must be due to redundancy in the angle deformation. After careful inspection it can be concluded that the redundancy relates to the $\angle FXeF$ angle deformations since, summation of all angles must be 360 degree, which is not possible in symmetric A_1 mode. This restriction does not apply to the four $\angle FXeO$ angle deformations.

Hence, finally we can assign the spectral observation in more details as shown in Table 7.21.

Table 7.21: Final assignment of spectral bands of $Xe(O)F_4$ molecule.

IR (vap, cm^{-1})	Raman(liquid, cm^{-1})	Assignment	Comments
926	920 (pol)	A_1; Xe−O stretching	Some points may be noted: * Stretching comes at higher energy than bending; compare 288 and 926 bands. * Xe−O and Xe−F both have A_1 symmetry, but Xe−O would be expected at higher energy as it is a stronger double bond. * The two deformations of E symmetry cannot be distinguished. * The two deformations of B_1 and B_2 symmetries cannot be distinguished. * Bending vibration involving two equatorial bonds is found to be at lower frequency than bending involving one axial and one equatorial bond.
609	605(depol)	E; Xe−F stretching	
576	567 (pol)	A_1; Xe−F stretching	
	527 (depol)	B_1; Xe−F stretching	
361	364 (depol)	E; F−Xe−O or F−Xe−F deformation	
288	286 (pol)	A_1; F−Xe−O deformation	
	232 (depol)	B_1 or B_2; F−Xe−O or F−Xe−F	
	220 (depol)	B_1 or B_2; F−Xe−O or F−Xe−F	
159	160 (depol)	E; F−Xe−O or F−Xe−F deformation	

7.3.5 $POCl_3$ (C_{3v} Point Group)

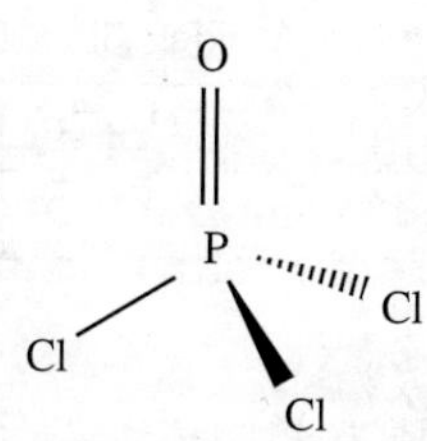

Taking 3N number of Cartesian coordinates as basis, we get the following reducible representation.

Table 7.22: Reducible representation taking cartesian coordinates as basis.

C_{3v}	E	$2C_3$	$3\sigma_v$
Γ_{3N}	15	0	3

After reduction, $\Gamma_{3N} = 4A_1 + A_2 + 5E$

Taking the help from character table, if we subtract the translations and rotations, we get

$$\Gamma_{vib} = 3A_1 + 3E$$

From character table, it can be concluded that both A_1 and E are IR as well as Raman active and also A_1 is Raman polarized and E is Raman depolarized.

Considering P−O bond vector, P−Cl bond vectors and bending modes for six bond angles ($\angle ClPCl$ and $\angle ClPO$) separately, we find the following reducible representations in Table 7.23.

Table 7.23: Reducible representations, along with their reduced forms, taking stretching and bending vectors.

C_{3v}	E	$2C_3$	$3\sigma_v$	After reduction
Γ_{P-O}	1	1	1	$= A_1$
Γ_{P-Cl}	3	0	1	$= A_1 + E$
$\Gamma_{bending}$	6	0	2	$= 2A_1 + E$

Hence, using internal coordinate method, if we add all the different symmetries, instead of earlier combination, $3A_1 + 3E$, here we get the total vibrations (Stretching and Bending) as, $\Gamma_{vib} = 4A_1 + 3E$. Here, one important observation is that there is one extra A_1 mode. This extra A_1 mode is supposed to arise due to redundancy. Since, during bending modes six angles are there which are supposed to move independently, but actually when five angles are changed, the sixth angle is automatically becomes fixed. Hence, this extra A_1 mode can safely be neglected and the correct symmetries of bending modes are:

$$\Gamma_{bending} = A_1 + 2E.$$

It is known that Raman polarized bands always arise from the totally symmetric vibrational mode, *i.e.*, A_1 mode. Therefore, the bands at 1290, 486 and 267 cm^{-1} are of A_1 type and remaining bands at 581, 337 and 193 cm^{-1} are of E type.

Keeping all those prior information and based on the above arguments we can continue the assignments of bands as shown in Table 7.24.

7.3.6 $[PtCl_4]^{2-}$ (D_{4h} Point Group)

The vibrational spectrum for this ion is given in the table below in Table 7.25.

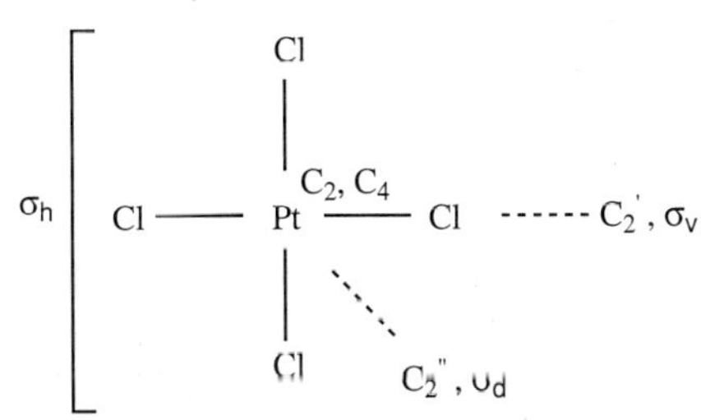

Table 7.24: Assignments of spectral bands of $POCl_3$ molecule.

IR (vap, cm^{-1})	**Raman(liquid, cm^{-1})**	**Assignment**	**Comments**
1297	1290 (pol)	A_1; P−O stretching (ν_1)	Some points may be noted: * Stretchings have higher frequency that bendings. * Double bonds have higher frequency than similar single bond. * Frequency generally decreases with increasing atomic mass. * First symmetric modes are taken and arranged in decreasing order and then number them. Then consider the other modes of lower symmetry.
580	581(depol)	E; P−Cl antisymmetric	
487	486 (pol)	A_1; P−Cl stretching (ν_2)	
340	337 (depol)	E; Antisymmetric bending (ν_5)	
267	267 (pol)	A_1; Symmetric bending (ν_3)	
Not accessible	193 (depol)	E; Antisymmetric bending (ν_6)	

Table 7.25: Experimental IR and Raman data of $[PtCl_4]^{2-}$ molecule.

IR (cm^{-1})	**Raman(cm^{-1})**
	332 (pol)
320	
	314 (depol)
183	
	170 (depol)
93	

Using the Cartesian coordinates for five atoms of the molecule as basis, the reducible representation is shown in Table 7.26.

Table 7.26: Reducible representation using cartesian coordinates as basis.

D_{4h}	E	$2C_4$	$C_2(=C_4^2)$	$2C_2'$	$2C_2''$	i	$2S_4$	σ_h	$2\sigma_v$	$2\sigma_d$
Γ	15	1	−1	−3	−1	−3	−1	5	3	1

Reduction formula and character table for D_{4h}, gives us:

$$\Gamma_{Total} = A_{1g} + A_{2g} + B_{1g} + B_{2g} + E_g + 2A_{2u} + B_{2u} + 3E_u$$

Among these A_{2u} and E_u is due to translation and A_{2g} and E_g is due to rotation.

Thus, the symmetries for the remaining nine vibrational motions are:

$$\Gamma_{vib} = A_{1g} + B_{1g} + B_{2g} + A_{2u} + B_{2u} + 2E_u$$

From the character table, as we have done in previous cases, among the nine vibrational motions, the modes corresponding to the symmetries of A_{2u} and E_u are IR active, and the modes corresponding to the symmetries of A_{1g}, B_{1g} and B_{2g} are Raman active. B_{2u} is a silent mode, it neither IR nor Raman active.

To go into further details of different vibrations and to find their symmetry labels, let us take the four Pt-Cl bond vectors as basis for construction of reducible representation for bond stretching modes and are presented in Table 7.27.

Table 7.27: Reducible representation taking the four Pt–Cl bond vectors as basis.

D_{4h}	E	$2C_4$	$C_2(=C_4^2)$	$2C_2'$	$2C_2''$	i	$2S_4$	σ_h	$2\sigma_v$	$2\sigma_d$
$\Gamma_{\text{Pt-Cl}}$	4	0	0	2	0	0	0	4	2	0

Here, it may be noted that unlike the construction of reducible representation taking three Cartesian coordinates as basis, here we do not count –1 with changing the direction of bond vectors, but consider the phase change. If we consider the expansion of bonds as positive phase then contraction would be negative phase. If phase changes during any symmetry operation, then only –1 would be counted.

Using reduction formula, we get

$$\Gamma_{\text{Pt-Cl}} = A_{1g} + B_{1g} + E_u$$

Therefore, we have four different Pt–Cl bond stretching vibrational modes and out of which one is totally symmetric mode (A_{1g}) and that must be Raman polarized.

Now, we have to identify bending or different deformation modes as well. They are of two types: one is *in-plane* mode and another one *out-of-plane* mode as shown in Figure 7.5.

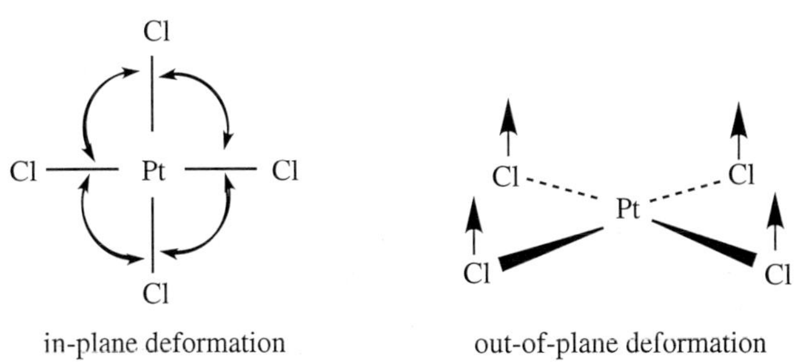

Figure 7.5: *in-plane* mode and *out-of-plane* mode of bending vibrations.

Double headed arrows are used for in-plane deformations and single headed arrows are used for designating out-of-plane modes. But, it is to be noted that as in the later case, the atoms can move in both + and – directions, hence reversal of the vectors can contribute –1 to the character, along with other characters of +1 and 0. The reducible representation for in-plane bending mode is shown in Table 7.28.

Table 7.28: Reducible representation corresponding to *in-plane* bending.

D_{4h}	E	$2C_4$	$C_2(=C_4^2)$	$2C_2'$	$2C''2$	i	$2S_4$	σ_h	$2\sigma_v$	$2\sigma_d$
$\Gamma_{\text{in-plane}}$	4	0	0	0	2	0	0	4	0	2

After reduction, we get

$$\Gamma_{\textit{in-plane}} = A_{1g} + B_{2g} + E_u$$

Here, one very interesting point is to be noted that one totally symmetric mode, A_{1g} is obtained. But careful inspection reveals that such kind of in-plane bond angle deformation practically is not feasible. This is because of the fact that simultaneous expansion or simultaneous contraction of all bond angles is not possible, as total angle of all bonds must be 360 degree. Hence, when three bonds expand, fourth bond must contract to keep total angle 360 degree. Thus, this kind of totally symmetric mode, A_{1g} is considered as redundant and can easily be discarded. Thus effectively , the *in-plane* contribution would be:

$$\Gamma_{\textit{in-plane}} = B_{2g} + E_u$$

At this point, our common sense finds the symmetry of *out-of-plane* deformations as:

$$\begin{aligned}\Gamma_{\textit{out-of-plane}} &= \Gamma_{\text{Vib}} - \left(\Gamma_{\text{Pt–Cl}} + \Gamma_{\textit{in-plane}}\right)\\ &= A_{1g} + B_{1g} + B_{2g} + A_{2u} + B_{2u} + 2E_u - (A_{1g} + B_{1g} + E_u + B_{2g} + E_u)\\ &= A_{2u} + B_{2u}\end{aligned}$$

In another way, taking the one headed arrows as basis, we can also find the symmetries of out-of-plane vibrations. The reducible representation corresponding to *out-of-plane* mode of vibrations is shown in Table 7.29.

Table 7.29: Reducible representation corresponding to *out-of-plane* mode of vibrations.

D_{4h}	E	$2C_4$	$C_2(=C_4^2)$	$2C_2'$	$2C_2''$	i	$2S_4$	σ_h	$2\sigma_v$	$2\sigma_d$
$\Gamma_{\text{out-of-plane}}$	4	0	0	–2	0	0	0	–4	2	0

The representation reduces to, $\Gamma_{out\text{-}of\text{-}plane} = E_g + A_{2u} + B_{2u}$

Here E_g is considered as redundant since, retaining its inversion center for an out-of-plane mode is quite impossible. Therefore, finally it becomes:

$$\Gamma_{out\text{-}of\text{-}plane} = A_{2u} + B_{2u}$$

which we have got earlier.

Thus, in a tabular form the final assigments can be presented as:

Table 7.30: Final assigments of bands for $[PtCl_4]^{2-}$ molecule.

IR (vap, cm^{-1})	**Raman (liquid, cm^{-1})**	**Assignment**	**Comments**
	332 (pol)	A_{1g}; Pt−Cl symmetric stretch	Some points may be noted: * Only totally symmetrical mode A_{1g} will be Raman polarized. * Bending comes at lower frequency than stretching. * In-plane bending comes at higher frequency than out-of-plane bending.
320		E_u; Pt−Cl asymmetric stretch	
	314 (depol)	B_{1g}; Pt−Cl stretch	
183		E_u; in-plane bend	
	170 (depol)	B_{2g}; in-plane bend	
93		A_{2u}; out-of-plane bend	

EXERCISES

7.1 Assuming all the Cl–O bonds as equivalent, find the symmetries of all the vibrations and assign each band given in the table below for $[ClO_4]^-$ species.

IR (cm^{-1})	Raman (cm^{-1})
1110	1102 (depol)
	935 (pol)
625	628 (depol)
	462 (depol)

7.2 MoF_5 molecule has a trigonal bipyramidal structure in gas phase. Show that at least two vibrations correspond to doubly degenerate symmetries and also assign the symmetry of each vibration from the experimental IR and Raman spectral data provided in the table below.

IR (cm^{-1})	Raman (cm^{-1})
	746 (pol)
730	733 (depol)
	703(pol)
685	
500	
	440(depol)
240	239 (depol)
205	200(depol)

7.3 Find the symmetries of different vibrational modes of $[SO_3]^{2-}$ ion and also assign the experimental Raman bands found at 966 (cm $^{-1}$, pol), 934 (cm $^{-1}$, depol), 620 (cm $^{-1}$, pol) and 472 (cm $^{-1}$, depol).

7.4 Find the symmetries of all possible vibrational modes of *trans*-$C_2H_2F_2$ following internal coordinates.

7.5 Find the translational and rotational degrees of freedom for naphthalene molecule taking the help of character table.

7.6 What is *redundant* symmetry? Explain this taking one example of a tetrahedral molecule.

Chapter 8

Selection Rule for IR and Raman Spectroscopy: Activity of Overtones, Combination Bands Based on Symmetry

8.1 Introduction

We know that various normal modes of a molecule can be catalogued and analyzed in terms of their symmetry with respect to the overall molecular symmetry. However, the thing is that all the normal modes may not necessarily be observed in the IR or Raman spectra. Hence, we basically need to know the spectroscopic activity of the normal modes or also called the spectroscopic selection rules to determine or predict which normal modes are active (allowed) or inactive (forbidden). The spectroscopic activity of any normal mode can be analyzed in terms of symmetry of the wave functions involved in the transition. The basic quantum mechanical selection rule for any such transition is that *the vibrational transition will have a non-zero intensity (i.e., allowed) in IR or Raman spectrum if the appropriate transition moment is non-zero*. Normal modes that do not follow this criterion can not be observed in spectra.

8.2 IR Activity

For IR absorption to occur, the normal mode of vibration must have an oscillating molecular dipole moment with the same frequency as the oscillating electric field of radiation. The transition moment (M) for the fundamental of a normal mode (0,1) can be written as:

$$\mathbf{M}(0,1) = \langle \Psi_0 | \mu | \Psi_1 \rangle$$

Where, μ is the resultant of its dipole moment components, μ_x, μ_y and μ_z.

Therefore, contribution to the total transition moment due to each dipole component can be written as:

$$M_x(0,1) = \langle \Psi_0 | \mu_x | \Psi_1 \rangle$$

$$M_y(0,1) = \langle \Psi_0 | \mu_y | \Psi_1 \rangle$$

$$M_z(0,1) = \langle \Psi_0 | \mu_z | \Psi_1 \rangle$$

If any one of these components becomes non-zero, the entire transition moment will be non-zero, *i.e.*, the transition becomes allowed and is said to be polarized along that particular component direction. If all the three integrals become non-zero then the transition is said to be non-polarised.

It is known that if the integrand is even or symmetric in nature, the magnitude of integration will be non-vanishing or non-zero and if the integrand is odd or antisymmetric in nature, the integration will vanish or gives zero value.

Applying this knowledge to the present case, in terms of symmetry, if the integral for the transition moment transforms as totally symmetric irreducible representation, the vibrational transition will be IR active or transition becomes allowed. **In general, the ground state vibrational wave function, Ψ_0 is considered as totally symmetric in nature for all molecules (except free radicals) and the excited state wave function, Ψ_1 is considered as the symmetry of the corresponding normal mode of vibration**. Again, it is already known that *the direct product of totally symmetric representation with any non-totally symmetric representation produces that non-totally symmetric representation.*

e.g., for C_{2v} point group: $A_1 \times B_1 = B_1$

Therefore, here the direct product of Ψ_0 and Ψ_1 produces the symmetry of Ψ_1. However, from another properties of direct product, *the product of an irreducible representation with itself (i.e., with same symmetry) results or contains the totally symmetric representation.* Since, a dipole component can be represented by a vector, we find that the symmetry species of μ_x, μ_y and μ_z are same as the unit vector transformations x, y and z, as listed in the character table. Thus, if any component of μ belongs to the same symmetry as Ψ_1, the product $\mu\Psi_1$ will be totally symmetric and hence, the integral will be non-vanishing.

Therefore, *a normal mode belonging to the same symmetry species as any of the unit vector transformations x, y or z will be active in IR spectrum.*

In summary, for a vibrational transition to be allowed, the transition dipole moment integral, $\mathbf{M}(0,1) = \langle \Psi_0 | \mu | \Psi_1 \rangle$ must be nonzero and consequently, the integral must be even or symmetric. Based on our knowledge gathered so far, the following conclusions can be drawn:

(i) The symmetry of the *ground state* wave function, Ψ_0 is *totally symmetric* in nature.

(ii) The symmetry of the *excited state* wave function, Ψ_1 possess the symmetry of the normal mode of vibration.

(iii) From the preceding two criterion, to get an even or symmetric integral, the nature of symmetry of μ must be the same as Ψ_1.

Let us continue the discussion by taking a few examples:

8.2.1 H_2O Molecule (C_{2v} Point Group)

For this point group, we know that the vibrational modes have the following symmetry,

$$\Gamma_{vib} = 2A_1 + B_2$$

It has been learnt that the excited state vibrational wave function, Ψ_1 must belong to either A_1 or B_2 symmetry and we have to find out the spectroscopic transition activity of these modes. In C_{2v} point group, the totally symmetric representation is A_1. Also the ground state symmetry always corresponds to totally symmetric representation *i.e.*, $\Psi_0 \sim A_1$. From the character table, the three dipole moment components transform as follows:

$$\mu_x(\text{or } x) \sim B_1$$
$$\mu_y(\text{or } y) \sim B_2$$
$$\mu_z(\text{or } z) \sim A_1$$

Consequently, the three integrals for A_1 mode of vibration become:

$$M_x(0,1) = \langle\Psi_0|\mu_x|\Psi_1\rangle = \langle A_1 \times B_1 \times A_1\rangle = B_1 \tag{8.1}$$

$$M_y(0,1) = \langle\Psi_0|\mu_y|\Psi_1\rangle = \langle A_1 \times B_2 \times A_1\rangle = B_2 \tag{8.2}$$

$$M_z(0,1) = \langle\Psi_0|\mu_z|\Psi_1\rangle = \langle A_1 \times A_1 \times A_1\rangle = A_1 \tag{8.3}$$

Since Eq. (8.1) and (8.2) do not produce a totally symmetric representation A_1 hence, these integrals will give zero value, and consequently, the two transition moment components, $M_x(0,1)$ and $M_y(0,1)$ become zero. On the other hand, since, Eq. (8.3) contains the totally symmetric representation A_1, the integration will be nonzero. Since, out of the three integrals at least one integral produces nonzero value, ultimately the transition for the vibrational mode will be IR active. Again, since the z-component is only nonzero, the transition is said to be *z-polarized* which basically means that the molecule can absorb *z-polarized* light but not x- or y-polarized light. This is known to us that *z-polarized* light means that light whose electric vector oscillates in the z-direction. A *z-polarized* transition depicts that the molecule absorbs only *z-polarized* light during that particular mode of vibration or in other words, the dipole moment is oscillating along z-axis during A_1 symmetry mode of vibration.

Similarly, the activity of B_2 mode can also be checked. The three integrals are as follows:

$$M_x(0,1) = \langle\Psi_0|\mu_x|\Psi_1\rangle = \langle A_1 \times B_1 \times B_2\rangle = A_2 \tag{8.4}$$

$$M_y(0,1) = \langle\Psi_0|\mu_y|\Psi_1\rangle = \langle A_1 \times B_2 \times B_2\rangle = A_1 \tag{8.5}$$

$$M_z(0,1) = \langle\Psi_0|\mu_z|\Psi_1\rangle = \langle A_1 \times A_1 \times B_2\rangle = B_2 \tag{8.6}$$

Here, it is found that since, at least one component (y-component) produces totally symmetric representation, the integral will be nonzero and hence, the overall transition becomes allowed and it is *y-polarised* transition as well.

8.2.2 NH_3 Molecule (C_{3v} Point Group)

Other common examples of this group are $CHCl_3$ or $POCl_3$ *etc.* Taking three cartesian coordinates as basis for NH_3 molecule, we get the symmetries of vibrations as:

$$\Gamma_{vib} = 2A_1 + 2E$$

Using character table, it is observed that the three dipole moment components (x,y and z directions) transform as symmetry of E, E and A_1 respectively. Similar to the earlier example, if the activity of A_1 mode of vibration are checked, the integrations are:

$$M_x(0,1) = \langle \Psi_0 | \mu_x | \Psi_1 \rangle = \langle A_1 \times E \times A_1 \rangle = E \quad (8.7)$$

$$M_y(0,1) = \langle \Psi_0 | \mu_y | \Psi_1 \rangle = \langle A_1 \times E \times A_1 \rangle = E \quad (8.8)$$

$$M_z(0,1) = \langle \Psi_0 | \mu_z | \Psi_1 \rangle = \langle A_1 \times A_1 \times A_1 \rangle = \mathrm{A}_1 \quad (8.9)$$

Out of these three integrations, equation (8.9) contains totally symmetric representation, A_1. As a result, the whole integration becomes nonzero and turns the vibrational mode as IR active and *z-polarized* as well.

Similarly, the E-symmetry mode of vibration can be considered and its activity can be checked. For this, the following are the three integrations:

$$M_x(0,1) = \langle \Psi_0 | \mu_x | \Psi_1 \rangle = \langle A_1 \times E \times E \rangle = \mathrm{A}_1 + \mathrm{A}_2 + \mathrm{E} \quad (8.10)$$

$$M_y(0,1) = \langle \Psi_0 | \mu_y | \Psi_1 \rangle = \langle A_1 \times E \times E \rangle = \mathrm{A}_1 + \mathrm{A}_2 + \mathrm{E} \quad (8.11)$$

$$M_z(0,1) = \langle \Psi_0 | \mu_z | \Psi_1 \rangle = \langle A_1 \times A_1 \times E \rangle = E \quad (8.12)$$

Here, the interesting notable point is that in equations (8.10) and (8.11), the direct product in the integrals produce reducible representaions and on subsequent reduction, the two reducible representations generate combination of three irreducible representations (*viz.*, $\mathrm{A}_1, \mathrm{A}_2$, E). Since, in the combination, totally symmetric representation, A_1 is present in both the equations, hence the transition dipole moment would be nonzero and therefore, the transition will be IR active and *xy-polarized* as well.

8.2.3 N_2 Molecule ($D_{\infty h}$ Point Group)

N_2 is a homonuclear diatomic molecule. From the character table, it is evident that the single stretching mode in N_2-molecule present is of totally symmetric, σ_g^+ type. A symmetric stretching vibrational mode must be present in any diatomic molecule. ψ_0 is known to be always totally symmetric, *i.e.*, σ_g^+ in nature . From the character table of ($D_{\infty h}$), the three components are found to be transformed as π_u (for x and y components) and σ_u^+ (for z component) symmetries. To check the IR activity of fundamental transition of stretching vibration in N_2 molecule, the transition moment integrals are as follows:

$$M_x(0,1) = \langle\Psi_0|\mu_x|\Psi_1\rangle = \langle\sigma_g^+ \times \pi_u \times \sigma_g^+\rangle = \pi_u$$

$$M_y(0,1) = \langle\Psi_0|\mu_y|\Psi_1\rangle = \langle\sigma_g^+ \times \pi_u \times \sigma_g^+\rangle = \pi_u$$

$$M_z(0,1) = \langle\Psi_0|\mu_z|\Psi_1\rangle = \langle\sigma_g^+ \times \sigma_u^+ \times \sigma_g^+\rangle = \sigma_u^+$$

Since, none of the triple products produce the totally symmetric representation, σ_g^+ hence, the magnitute of transition dipole moment integral is necessarily be equal to zero. Therefore, transition due to stretching vibration of N_2 is known to be forbidden and will not appear in infrared spectrum. Similar observations are found in case of any homonuclear diatomic molecule and consequently they are known as infrared inactive.

8.2.4 Carbon Monoxide Molecule ($C_{\infty v}$ Point Group)

From the character table, the only stretching mode of $C_{\infty v}$ point group present is of σ^+ symmetry. A symmetric stretching vibrational mode must be present in any diatomic molecule. ψ_0 is always totally symmetric in nature, $i,e.,\sigma^+$. Dipole moment operator has three components along three coordinates x,y and z. From character table, the three components transform as π (for x and y components) and σ^+ (for z component) .To check the IR activity of the fundamental transition of CO molecule, the transition moment integrals are as follows:

$$M_x(0,1) = \langle\Psi_0|\mu_x|\Psi_1\rangle = \langle\sigma^+ \times \pi \times \sigma^+\rangle = \pi$$

$$M_y(0,1) = \langle\Psi_0|\mu_y|\Psi_1\rangle = \langle\sigma^+ \times \pi \times \sigma^+\rangle = \pi$$

$$M_z(0,1) = \langle\Psi_0|\mu_z|\Psi_1\rangle = \langle\sigma^+ \times \sigma^+ \times \sigma^+\rangle = \sigma^+$$

Since, the direct product produces the totally symmetric representation, σ^+, hence the transition is IR active.

8.3 Raman Activity

For Raman activity, there should be a change in the polarisibilty (α) with the normal coordinate at the equilibrium configuration, *i.e.*, $(d\alpha/dQ)_{eq} \neq 0$.
Where, P= αE (P is the induced moment and E is the strength of electric field).
The polarisibility is a tensor quantity and can be best expressed as:

$$\begin{bmatrix} Px \\ Py \\ Pz \end{bmatrix} = \begin{bmatrix} \alpha_{xx} & \alpha_{xy} & \alpha_{xz} \\ \alpha_{yx} & \alpha_{yy} & \alpha_{yz} \\ \alpha_{zx} & \alpha_{zy} & \alpha_{zz} \end{bmatrix} . \begin{bmatrix} Ex \\ Ey \\ Ez \end{bmatrix}$$

Where, $\alpha_{ij} = \alpha ji$

If the change in any one of these components is nonzero, $[(d\alpha_{ij}/dQ)_{eq} \neq 0]$ then a vibrational mode will become Raman active. In Raman spectroscopy, transition between vibrational states can be described in a manner similar to that of IR spectroscopy as:

$$\langle \Psi_i|\alpha|\Psi_j\rangle$$

As described earlier α is a tensor with 3×3 array of components. Now, since $\alpha_{ij} = \alpha_{ji}$, hence, out of total nine terms, effectively only six terms are considered. Hence, similar to dipole moment in IR spectroscopy, here also we have to consider total six integrals corresponding to six α's. For any vibrational mode to be Raman active, out of the six, we need at least one integral to be non-zero.

Let us take few examples and discuss the Raman activity.

8.3.1 H_2O Molecule (C_{2v} Point Group)

We have already found the symmetries of various vibrations of H_2O molecule as:

$$\Gamma_{Vib} = 2A_1 + B_2$$

As it is already mentioned that for any vibrational mode, the ground state always corresponds to totally symmetric representation and the excited state parallels the symmetry of that particular vibrational mode. Hence, for the integral $\langle \Psi_i|\alpha|\Psi_j\rangle$, the symmetry of Ψ_0 is A_1 for C_{2v} point group molecule. The character table, shown below, designates the symmetries of α_{xy} and α_{yx} as A_2; α_{yz} and α_{zy} as B_2; α_{xz} and α_{zx} as B_1. Also, the symmetries of the another three polarisibility components, α_{xx}, α_{yy} and α_{zz} carry the same symmetry and corresponds to A_1.

Table 8.1: Character Table of the C_{2v} Point Group.

C_{2v}	E	C_2	$\sigma_v(xz)$	$\sigma_v(yz)$	Symmetry of functions
A_1	1	1	1	1	$z,\ (x^2, y^2, z^2)$
A_2	1	1	–1	1	R_x, xy
B_1	1	–1	1	–1	x, R_y, xz
B_2	1	–1	–1	1	y, R_x, yz

Now, let us check the Raman activity of two **A_1 symmetry** mode vibrations. For that we need to find the direct products corresponding to six integrals.

$$\langle \Psi_i|\alpha_{xx}|\Psi_j\rangle = \langle A_1 \times A_1 \times A_1\rangle = A_1 \qquad \langle \Psi_i|\alpha_{xy=yx}|\Psi_j\rangle = \langle A_1 \times A_2 \times A_1\rangle = A_2$$

$$\langle \Psi_i|\alpha_{yy}|\Psi_j\rangle = \langle A_1 \times A_1 \times A_1\rangle = A_1 \qquad \langle \Psi_i|\alpha_{yz=zy}|\Psi_j\rangle = \langle A_1 \times B_2 \times A_1\rangle = B_2$$

$$\langle \Psi_i|\alpha_{zz}|\Psi_j\rangle = \langle A_1 \times A_1 \times A_1\rangle = A_1 \qquad \langle \Psi_i|\alpha_{zx=xz}|\Psi_j\rangle = \langle A_1 \times B_1 \times A_1\rangle = B_1$$

We know that the integral which contains totally symmetric (or even) function will give non-zero value and other integrals will vanish. Since, at least few integrals result non-zero value, hence the A_1 type of vibrational mode will be Raman active. Also in Raman spectroscopy a symmetrical vibrational mode, on incident upon a polarized light, results with a polarized light as well, as exiting radiation.

Similarly, we can check the Raman activity of the **B_2 mode** of vibration:

$$\langle\Psi_i|\alpha_{xx}|\Psi_j\rangle = \langle A_1\times A_1\times B_2\rangle = B_2 \qquad \langle\Psi_i|\alpha_{xy=yx}|\Psi_j\rangle = \langle A_1\times A_2\times B_2\rangle = B_1$$

$$\langle\Psi_i|\alpha_{yy}|\Psi_j\rangle = \langle A_1\times A_1\times B_2\rangle = B_2 \qquad \langle\Psi_i|\alpha_{yz=zy}|\Psi_j\rangle = \langle A_1\times B_2\times B_2\rangle = A_1$$

$$\langle\Psi_i|\alpha_{zz}|\Psi_j\rangle = \langle A_1\times A_1\times B_2\rangle = B_2 \qquad \langle\Psi_i|\alpha_{zx=xz}|\Psi_j\rangle = \langle A_1\times B_1\times B_2\rangle = A_2$$

Since, at least one integral give rise to totally symmetric A_1 representation hence, the whole integral becomes non-zero and hence B_2 symmetry mode of vibration is said to be Raman active. Again as this mode of vibration is asymmetric in nature, the emitted light is ended up with a depolarized light, even if the incident light is polarized in nature.

8.3.2 NH_3 Molecule (C_{3v} Point Group)

This is already known that the symmetries of various vibrations of NH_3 molecule are found to be as:

$$\Gamma_{Vib} = 2A_1 + 2E$$

This equation indicates that there are two vibrations of A_1 symmetry and another two triply degenerate vibrations are of E symmetry. From the character table of C_{3v}, shown below, the symmetries of various polaribility components are as follows: $\alpha_{xy} = \alpha_{yx}$, $\alpha_{yz} = \alpha_{zy}$ and $\alpha_{xz} = \alpha_{zx}$ corresponds to A_1 symmetry; another three components, α_{xx} , α_{yy} , α_{zz} correspond to E symmetry. Also this is known, that the symmetry of final state, Ψ_j is either of A_1 or E symmetry and that of initial state, Ψ_i is always of A_1 symmetry.

Table 8.2: Character table for the C_{3v} point group.

C_{3v}	E	$2C_3$	$3\sigma_v$	Symmetry of function
A_1	1	1	1	$z,\ (x^2+y^2), z^2$
A_2	1	1	–1	R_z
E	2	–1	0	$(x,y)(R_x,R_y),(x^2+y^2,xy)(yz,xz)$

Now, we have to check the various vibrational modes corresponding to the two symmetries.

For A_1 mode:

$$\langle\Psi_i|\alpha_{xx}|\Psi_j\rangle = \langle A_1\times A_1\times A_1\rangle = A_1 \qquad \langle\Psi_i|\alpha_{xy=yx}|\Psi_j\rangle = \langle A_1\times E\times A_1\rangle = E$$

$$\langle\Psi_i|\alpha_{yy}|\Psi_j\rangle = \langle A_1\times A_1\times A_1\rangle = A_1 \qquad \langle\Psi_i|\alpha_{yz=zy}|\Psi_j\rangle = \langle A_1\times E\times A_1\rangle = E$$

$$\langle\Psi_i|\alpha_{zz}|\Psi_j\rangle = \langle A_1\times A_1\times A_1\rangle = A_1 \qquad \langle\Psi_i|\alpha_{zx=xz}|\Psi_j\rangle = \langle A_1\times E\times A_1\rangle = E$$

For E mode:

($\alpha's$ corresponding to A_1)	($\alpha's$ corresponding to E)
$\langle A_1\times A_1\times E\rangle = E$	$\langle A_1\times E\times E\rangle = A_1+A_2+E$
$\langle A_1\times A_1\times E\rangle = E$	$\langle A_1\times E\times E\rangle = A_1+A_2+E$
$\langle A_1\times A_1\times E\rangle = E$	$\langle A_1\times E\times E\rangle = A_1+A_2+E$

In both the vibrational modes, the direct product of at least one integral is found to result totally symmetric A_1 or resultant combination contains the A_1 symmetry. Therefore, both the vibrational modes are Raman active, and the symmetric vibrational mode, A_1 results in polarized radiation, whereas the asymmetric mode, E results in a depolarized radiation.

8.3.3 CH_4 Molecule (T_d Point Group)

This is already known that the symmetries of various vibrations of CH_4 molecule can be expressed as:

$$\Gamma_{Vib} = A_1 + E + 2T_2$$

We can see that there are total nine vibrations and out of those, two vibrations are triply degenerate, one is doubly degenerate and another one is totally symmetric in nature.

It may be recalled that Ψ_i is always of A_1 symmetry and the symmetry of Ψ_j will be according to the corresponding symmetries of vibrations, *i.e.*, either A_1 or E or T_2 depending upon the nature of a particular vibrational mode.

From the character table of T_d, shown below, the symmetries of various polarizability components are as follows:

Table 8.3: Character Table of the T_d Point Group.

T_d	E	$8C_3$	$2C_2$	$6S_4$	$6\sigma_d$	Symmetry of functions
A_1	1	1	1	1	1	$x^2+y^2+z^2$
A_2	1	1	–1	–1	–1	
E	2	–1	2	0	0	$(2z^2-x^2-y^2, x^2-y^2)$
T_1	3	0	–1	1	–1	R_x, R_y, R_z
T_2	3	0	–1	–1	1	$(x,y,z)(xy,yz,xz)$

If we look at the character table, it is found that the symmetries of α's are as follows:
The components α_{xy} (or, α_{yx}) , α_{yz} (or, α_{zy}) and α_{xz} (or, α_{zx}) corresponds to T_2 symmetry and $\alpha_{xx}, \alpha_{yy}, \alpha_{zz}$ polarisibility components correspond to both A_1 and E symmetries.

Hence, in this case, for α_{xx}, α_{yy} and α_{zz} polarisibility components, along with A_1-symmetry, we have to find another three extra integrals corresponding to E-symmetry as well, for each vibrational mode. Hence, the integrals, $\langle\Psi_i|\alpha's|\Psi_j\rangle$ for the three modes of vibrations are as follows:

For A_1 mode:

($\alpha's$ corresponding to A_1)	($\alpha's$ corresponding to E)	($\alpha's$ corresponding to T_2)
$\langle A_1 \times A_1 \times A_1\rangle = A_1$	$\langle A_1 \times E \times A_1\rangle = E$	$\langle A_1 \times T_2 \times A_1\rangle = T_2$
$\langle A_1 \times A_1 \times A_1\rangle = A_1$	$\langle A_1 \times E \times A_1\rangle = E$	$\langle A_1 \times T_2 \times A_1\rangle = T_2$
$\langle A_1 \times A_1 \times A_1\rangle = A_1$	$\langle A_1 \times E \times A_1\rangle = E$	$\langle A_1 \times T_2 \times A_1\rangle = T_2$

For E mode:

($\alpha's$ corresponding to A_1)	($\alpha's$ corresponding to E)	($\alpha's$ corresponding to T_2)
$\langle A_1 \times A_1 \times E \rangle = E$	$\langle A_1 \times E \times E \rangle = A_1 + A_2 + E$	$\langle A_1 \times T_2 \times E \rangle = T_1 + T_2$
$\langle A_1 \times A_1 \times E \rangle = A_1$	$\langle A_1 \times E \times E \rangle = A_1 + A_2 + E$	$\langle A_1 \times T_2 \times E \rangle = T_1 + T_2$
$\langle A_1 \times A_1 \times E \rangle = A_1$	$\langle A_1 \times E \times E \rangle = A_1 + A_2 + E$	$\langle A_1 \times T_2 \times E \rangle = T_1 + T_2$

For T_2 mode:

($\alpha's$ corresponding to A_1)	($\alpha's$ corresponding to E)	($\alpha's$ corresponding to T_2)
$\langle A_1 \times A_1 \times T_2 \rangle = T_2$	$\langle A_1 \times E \times T_2 \rangle = T_1 + T_2$	$\langle A_1 \times T_2 \times T_2 \rangle = A_1 + E + T_1 + T_2$
$\langle A_1 \times A_1 \times T_2 \rangle = T_2$	$\langle A_1 \times E \times T_2 \rangle = T_1 + T_2$	$\langle A_1 \times T_2 \times T_2 \rangle = A_1 + E + T_1 + T_2$
$\langle A_1 \times A_1 \times T_2 \rangle = T_2$	$\langle A_1 \times E \times T_2 \rangle = T_1 + T_2$	$\langle A_1 \times T_2 \times T_2 \rangle = A_1 + E + T_1 + T_2$

It is found that the integrals corresponding to all the three modes of vibrations, either results totally symmetric representation, A_1 or contains totally symmetric representation, A_1 as one of the component. Therefore, it can be concluded that the three modes of vibrations are Raman active.

8.4 Spectroscopic Activity of Overtones, Combination Bands Based on Symmetry

We have learnt how to predict the number of the fundamental transitions of normal modes of vibrations and their spectroscopic activities . However, in most of the cases, these predictions are far from actual experimentally observed spectra. Weak intensities or instrumental limitations can sometimes result in fewer observable frequencies than predicted. But in most of the cases, there appears more number peaks in the spectra than predicted. Several factors can give rise to additional peaks in the IR and Raman spectra, the most common being the presence of overtone bands and combination bands.

It is known that the selection rule for IR transition is $\Delta v = \pm 1$ for harmonic oscillator, and $\Delta v = \pm 1, \pm 2, \pm 3...$*etc*., due to anharmonicity in the oscillation and these later transitions are the origin of appearance of *overtones*. For a normal mode ν_i, the overtones appear approximately at frequency of $2\nu_i$, $3\nu_i$ *etc*. There is a possibility that two separate normal modes may couple to give rise to *combination bands* and *difference bands* at $\nu_i + \nu_k$ and $\nu_i - \nu_k$respectively and consequently they also may have their respective overtone bands, although intensities would be very weak.

We have learnt that for fundamental transition to be IR allowed the symmetry of the normal mode must belong to the same symmetry or irreducible representation as one or more components of the dipole moment vector, which is equivalent to the symmetry species of the unit vectors (*i.e.* x, y or z) listed in the character table.

Similarly, to be Raman active, the symmetry of the normal mode must belong to the same symmetry or irreducible representation as one or more components of polarisibility tensor, which is equivalent to the symmetry species of the binary products ($i.e., x^2, y^2, xy, x^2 - y^2 etc.$) listed in the character table.

The same general rules are also applicable for overtones and combination bands. The symmetries of overtones and combination transitions can be determined by taking the direct products of the irreducible representations of the fundamentals involved. If we recall the properties of direct product, we know that the direct product of any nondegenerate irreducible representation with itself is the totally symmetric representation. Hence, it may be concluded that the first overtone ($\Gamma_i \times \Gamma_i$ = always totally symmetric representation) of any nondegenerate normal modes will belong to the totally symmetric representation. From character table it is evident that totally symmetric representation is always the symmetry species of at least one component of the polarizibility tensor. Therefore, it can be concluded that *all first overtones of nondegenerate normal modes will be Raman allowed.*

Since, direct product of any irreducible representation or symmetry with itself (as in case of 1st overtone) produces the totally symmetric representation, further careful inspection of character table reveals that first overtones of molecules of most of the point groups will be IR inactive, except for molecules belonging to the point groups $C_1, C_s, C_n,$ and C_{nv}. The exception arises in these four point groups since, the totally symmetric representation contains at least one component of unit vectors (*i.e.* x, y or z present) in the character tables of those point groups. In any other point group, the totally symmetric representation does not contain any component of dipole moment vector (*i.e.*, x or y or z).

We can take the example of NH_3 (C_{3v}) molecule to elaborate the above discussion. The character table is presented below again for our reference. Let us take

Table 8.4: Character table for the C_{3v} point group.

C_{3v}	E	$2C_3$	$3\sigma_v$	Symmetry of function
A_1	1	1	1	$z, (x^2+y^2), z^2$
A_2	1	1	−1	R_z
E	2	−1	0	$(x,y)(R_x,R_y),(x^2+y^2,xy)(yz,xz)$

the fundamental of A_2 symmetry IR-inactive vibration and check the activity of first overtones. For that we need to calculate the direct product, $A_2 \times A_2$ (Since, overtone is $2\nu_i$):

$$A_2 \times A_2 = A_1$$

This result is obvious from direct product rule (Direct product of any irreducible representation with itself produces totally symmetric representation). The A_1 species does contain one component of electric dipole (here z component), cited in the character table. Hence, the first overtone of fundamental IR-inactive A_2 mode is found to be IR as well as Raman active. It is worthless to mention again that first overtone of any symmetry species is Raman active since, totally symmetric irreducible representation always found to contain quadratic terms as evident from any point group character table. The discussion can be extended by considering the 2nd overtone of A_2 mode. Here, we need to find the direct product, $A_2 \times A_2 \times A_2$ (as 2nd overtone is $3\nu_i$) and the result is found to be A_2 which is a clear indication of IR inactivity.

It is well known that the direct product of any **degenerate representation** with itself contains the totally symmetric representation along with other irreducible representations. A tetrahedral molecule can be taken as an example to illustrate the discussion for degenerate representation.

Table 8.5: Character Table of the T_d Point Group.

T_d	E	$8c_3$	$2c_2$	$6S_4$	$6\sigma_d$	Symmetry of functions
A_1	1	1	1	1	1	$x^2+y^2+z^2$
A_2	1	1	−1	−1	−1	
E	2	−1	2	0	0	$(2z^2-x^-y^2, x^2-y^2)$
T_1	3	0	−1	1	−1	R_x, R_y, R_z
T_2	3	0	−1	−1	1	$(x,y,z)(xy,yz,xz)$

Let us consider the possible activities of first overtone of T_2 symmetry (ν_3 frequency of CH_4) whose fundamental mode is IR as well as Raman active. To find out first overtone ($2\nu_3$) we need to carry out the direct product, $(T_2 \times T_2)$. The characters of reducible representation, thus formed, are found to be as $\Gamma = 9, 0, 1, 1, 1$ and after reduction we get,

$$\Gamma = A_1 + E + T_1 + T_2$$

Since, it contains A_1, E and T_2 , where polarisibility components are present hence, first overtone is Raman active. Again, due to the presence of dipole moment componets in T_2 mode, as evident from character table, makes this mode IR active as well. Therefore, the conclusion is that the first overtone of vibrational mode of CH_4 associated with T_2 symmetry is Raman as well as IR active. The activities of combinations of frequencies can also be determined in a similar manner. Considering the combination, ν_2 (1526 cm^{-1}) + ν_4 (1306 cm^{-1}) of CH_4 corresponding to E and T_2 symmetry respectively, for two fundamental modes, the direct product is found as: $(E \times T_2) = 6, 0, -2, 0, 0$ which subsequently reduces to $T_1 + T_2$. Since, it contains or associated to T_2 symmetry (where both dipole moment components as well as polarisibility components are present) hence, this combination transition would be both IR and Raman active. An IR band at 2823 cm^{-1} has been assigned to this combination (1526 + 1306 = 2832 cm^{-1}). The difference bands are also determined in the same way as the positive combination has been done.

8.5 Fermi Resonance

Although generally weak, overtones and combinations bands can show sometimes , surprisingly strong intensities when their frequencies are very close to a fundamental frequency of other vibrational mode with a matching of symmetry. In these cases, generally, the overtone or combination band "borrows" intensity from the fundamental. The extent of mixing is inversely proportional to the difference in energy of the two participating states. The two bands get mixed as well as shows a splitting in energy levels, losing their individual identities. One band moves to higher frequency and the other to lower, with both are having comparable intensities.

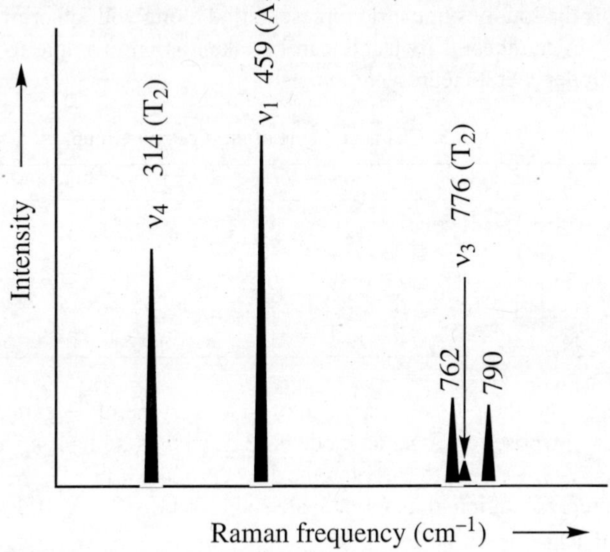

Figure 8.1: Qualitative vibrational Raman spectrum of liquid CCl_4.

This phenomenon is known as *Fermi resonance*. Fermi resonance is a necessary consequence of symmetry.

A well-known case of Fermi resonance occurs in the Raman spectrum of CCl_4. It has fundamentals ν_1 of A_1 symmetry at 459 cm^{-1}, ν_3 of T_2 symmetry at 776 cm^{-1} and ν_4 of T_2 symmetry at 314 cm^{-1}. The combination $\nu_1 + \nu_4$ has T_2 symmetry ($A_1 \times T_2 = T_2$) and expected to appear at 773 cm^{-1} (combination of 459 and 314 cm^{-1}) , which is virtually almost same as ν_3. Due to same symmetry (T_2) as well as energy matching of ν_3 mode and the combination mode, ($\nu_1 + \nu_4$) can interact and get mixed and consequently due to this Fermi resonance, in actual observed spectra two peaks appear at 762 cm^{-1} and 790 cm^{-1}. As shown in Figure 8.1, neither of these bands could be assigned exclusively to ν_3 or ($\nu_1 + \nu_4$).

8.6 Vibrational Activities of Linear Molecules

Linear molecules (Point group $C_{\infty v}$ and $D_{\infty h}$) involve infinite number of symmetry operations. It is not possible to resolve a reducible representation into its corresponding irreducible representations due to the presence of infinity order, $h = \infty$ for those groups. It is therefore, necessary to carry out the resolution by alternative inspection method. Taking the example of a linear molecule, CO_2 with point group $D_{\infty h}$ with the molecular axis along the bond as the z-axis, the reducible representation, based on rotating the molecule about the z-axis by an angle of ϕ, the following table can be constructed.

Table 8.6: Table to calculate reducible representaation.

$(D_{\infty h})$	E	$2C_{\infty}(\phi)$	....$\infty\sigma_v$	i	$2S_{\infty}$	.......∞C_2
Number of unshifted atoms	3	3	3	1	1	1
Γ	9	3(1+2cosϕ)*	3	–3	(–1+2cosϕ)	–1

*The z-vector contributes 1 as it remains unshifted. Each of x and y-vectors contributes cos ϕ as the component of displaced vector along the original vector is cos ϕ.

From character table of $(D_{\infty h})$ it is evident that the contribution from translational and rotation are as follows:

$(D_{\infty h})$	E $2C_{\infty}(\phi)$ .... $\infty\sigma_v$ i $2S_{\infty}$ .... ∞C_2
$\Gamma_{trans}+\Gamma_{rot}$	5 1+4cosϕ ..... 1 –1 –1 .... -1

Hence, the contribution from vibrational degrees of freedom would be:

$(D_{\infty h})$	E $2C_{\infty}(\phi)$ .... $\infty\sigma_v$ i $2S_{\infty}$ .... ∞C_2
$\Gamma_{trans}+\Gamma_{rot}$	4 2+2cosϕ .....2 –2 2cosϕ 0

After reduction of this reducible representation, we get the following combination for Γ_{vib}:

$$\Gamma_{vib} = \Sigma_g^+ + \Sigma_u^+ + \Pi_u$$

Thus, there are two nondegnerate vibrations (Σ_g^+ and Σ_u^+ one of which is totally symmetric) and two doubly-degenerate vibrations. Again, the reducible representation is constructed based on the two C=O bond-stretching vectors, $\mathbf{r_1}$ and $\mathbf{r_2}$ as:

$(D_{\infty h})$	E $2C_{\infty}(\phi)$ .... $\infty\sigma_v$ i $2S_{\infty}$ .... ∞C_2
$\Gamma_{stretching}$	2 2 ... ...2 0 0 0

After reduction to its corresponding irreducible components,

$$\Gamma_{stretching} = \Sigma_g^+ + \Sigma_u^+$$

Thus, out of the three vibrational modes, the remaining bending vibration corresponds to the remaining symmetry, Π_u. Since, CO_2 is a centrosymmetric molecule, the rule of Mutual Exclusion is followed by this molecule. From the inspection of the character table, the presence of bases in the last columns of character table, depicts that the vibration corresponding to Σ_g^+ symmetry is IR inactive and Raman active, whereas, the vibrations corresponding to Σ_u^+ and Π_u symmetries are IR active but Raman inactive.

EXERCISES

8.1 Find the symmetries and polarization of transitions corresponding to the allowed IR bands of water molecule.

8.2 The IR spectrum of CH_4 molecule shows two bands at 3020 cm^{-1} and 1306 cm^{-1} of T_2 symmetry. Two relatively weak bands are observed at 4313 cm^{-1} and 1720 cm^{-1}. Comment on the feasibility of assigning them as combination or difference bands of the initial two.

8.3 Determine the polarization of exiting Raman line corresponding to each allowed vibrational mode for a bend AB_2 type molecule.

8.4 How many vibrations are possible for a XY_3 type of molecule? Out of those how many will be IR or Raman active? Find their symmetries.

8.5 Discuss the selection rules for fundamental, overtone and combinationa bands of formaldehyde molecule.

8.6 What are the basic criteria for a molecule to be IR and Raman allowed in terms of symmetry point of view?

Chapter 9

Application of Symmetry in Electronic Transitions

9.1 Introduction

Electronic transition occurs from one atomic or molecular orbital to another atomic or molecular orbital in presence of ultraviolet or visible electromagnetic radiation. Although all the transitions are inherently equally probable, practically all the transitions are still not equally probable. Transition activities are guided by some selection rules which are basically consequence of the calculation of transition moment integral, M. We know that for an electronic transition, the dipole moment integral, $M = \langle\Psi_0|\mu|\Psi_1\rangle$ must be non-zero. Since, dipole moment vector has three components, μ_x, μ_y and μ_z hence, the overall integral looks like:

$$\langle\Psi_0 \begin{pmatrix} \mu_x \\ \mu_y \\ \mu_z \end{pmatrix} \Psi_1\rangle$$

It is known that each component of dipole moment operator has odd (u) parity. Hence, in order to get the total integral *even*, the direct product of the representations of the initial and final states must be *odd*. In other words, the parity of the initial and the final states must be of opposite or alternate nature.

As a consequence of this integral, the selection rule for electronic transition of atoms evolve as: *The electronic transitions between states of the same parity are forbidden, i.e., s-s, p-p or d-d* transitions are forbidden while *s-p, p-d or d-f* are allowed.

We have another selection rule as: *The spin of the initial and final states should be conserved.* Later the evolution of this rule would be discussed.

Now during the prediction of the feasibility of electronic transitions in a molecule, it is very much practical to work with the symmetries of the ground and excited electronic states of the molecule instead of the orbitals involved in the electronic transition. *The symmetry of electronic state of a molecule is determined by evaluating direct product of corresponding representations of the involved molecular orbitals. For singly-occupied molecular orbital, the symmetry is that of the molecular orbital itself (for single-electron orbital, a lower case Mulliken symbol is used. The symmetry of electronic state of the molecule is represented by Capital letter). For doubly-occupied molecular orbital, the direct product of the same two representations always yields a totally symmetric representation and thus may be ignored in evaluating the direct product of the state of the molecule.*

9.1.1 Selection Rules

For an electronic transition, the transition moment integral is defined as:

$$M = \langle \Psi_0 | \mu | \Psi_1 \rangle$$

(Ψ_0 and Ψ_1 are wave functions corresponding to initial and final states)

$$M = \langle \psi_{0(es)} \psi_{0(v)} | (\mu_n + \mu_e) | \psi_{1(es)} \psi_{1(v)} \rangle$$

$$[\text{Since}, \mu = \mu_n + \mu_e \text{ and } \psi = \psi_{(es)} \psi_{(v)}]$$

$$M = \langle \psi_{0(es)} \psi_{0(v)} | \mu_n | \psi_{1(es)} \psi_{1(v)} \rangle + \langle \psi_{0(es)} \psi_{0(v)} | \mu_e | \psi_{1(es)} \psi_{1(v)} \rangle$$

$$[\text{Following Born-Oppenheimer approximation}]$$

$$M = \langle \psi_{0(v)} | \mu_n | \psi_{1(v)} \rangle \langle \psi_{0(es)} | \psi_{1(es)} \rangle + \langle \psi_{0(es)} | \mu_e | \psi_{1(es)} \rangle \langle \psi_{0(v)} \psi_{1(v)} \rangle$$

In this equation, out of these two parts, since, $\langle \psi_{0(es)} | \psi_{1(es)} \rangle$ is zero (Two electronic wave functions are always orthogonal), hence, the whole 1st part results zero. In the 2nd part, the vibrational component, $\langle \psi_{0(v)} \psi_{1(v)} \rangle$ is not necessarily be zero since, two vibrational wave functions belong to two different electronic states (Hence, not orthogonal). This is known as *Franck-Condon* factor. Therefore,

$$M = \langle \psi_{0(es)} | \mu_e | \psi_{1(es)} \rangle \langle \psi_{0(v)} \psi_{1(v)} \rangle$$

Considering the spin components, M can further be splitted and can be written as:

$$M = \langle \psi_{0(e)} | \mu_e | \psi_{1(e)} \rangle \langle \psi_{0(v)} \psi_{1(v)} \rangle \langle \psi_{0(s)} \psi_{1(s)} \rangle$$

The 1st and 3rd integrals of this equation are the basis for *orbital* and *spin selection rule* respectively.

9.1.2 Orbital Selection Rule

We have already leart to get the symmetry of different occupied orbitals. Also, we need to remember that an electronic transition is orbitally allowed if and only if the triple direct product, $\Gamma_{\psi_{0(e)}} \times \Gamma_{(\mu_e)} \times \Gamma_{\psi_{1(e)}}$ becomes totally symmetric irreducible representation or contains totally symmetric irreducible representation. The dipole moment operator transform as x, y or z and found to be of σ_u^+ and π_u symmetry, say, for a linear molecule ($D_{\infty h}$ point group).

Let us consider a spin allowed (singlet to singlet) transition, ${}^1\Sigma_g^+ \rightarrow {}^1\Sigma_u^+$ for a linear molecule. We can find the triple direct product as:

$$\sigma_u^+ \times \sigma_u^+ \times \sigma_g^+ = \sigma_g^+ \text{ (corresponding to } \mu_z)$$

and

$$\sigma_u^+ \times \pi_u \times \sigma_g^+ = \pi_g \text{ (corresponding to } \mu_x \text{ and } \mu_y)$$

Since, one component of μ (specifically μ_z) produces totally symmetric IR, σ_g^+, hence this transition is orbitally allowed. Hence, this kind of transition is spin as well as orbitally allowed or also called fully allowed transition. Hence, this kind of transition will show intense peak in spectrum. We consider another transition which is spin forbidden but orbitally allowed, say, ${}^1\Sigma_g^+ \rightarrow {}^3\Sigma_u^+$. This transition will be again orbitally allowed as we have seen earlier, but since spin selection rule is the most dominating of all selection rules, this kind of transition will give very weak intensity.

9.1.3 Spin Selection Rule

Let us recall the equation for transition moment,

$$M = \langle \psi_{0(e)} | \mu_e | \psi_{1(e)} \rangle \langle \psi_{0(v)} \psi_{1(v)} \rangle \langle \psi_{0(s)} \psi_{1(s)} \rangle$$

As we know that a transition to be allowed, the whole integral must be nonzero. Let us consider a transition, which is orbitally and vibrationally allowed, *i.e.* first two integrals already result even or symmetric function. Hence, the overall integral to be symmetric, the direct product of the two spin states (initial, $\psi_{0(s)}$ and final, $\psi_{1(s)}$ spin states) of the 3rd integral must be symmetric. The direct product of the two spin states can only be symmetric if and only if they possess the same nature, *i.e.*, either both are even or symmetric or both are antisymmetric or odd in nature. Therefore, the spin selection rule can be designated as $\Delta S = 0$.

9.1.4 Vibronic Selection Rule

We have the equation, $M = \langle \psi_{0(e)} | \mu_e | \psi_{1(e)} \rangle \langle \psi_{0(v)} \psi_{1(v)} \rangle \langle \psi_{0(s)} \psi_{1(s)} \rangle$
Keeping the spin component silent, if we rearrange the equation, we get

$$M = \langle \psi_{0(e)} | \mu_e | \psi_{1(e)} \rangle \langle \psi_{0(v)} \psi_{1(v)} \rangle$$

This integral is the basis for vibronic selection rules and for a transition from initial state to a final state to be allowed this integral must be nonzero.

Since, the initial or ground vibrational state, $\psi_{0(v)}$ is known to be always totally symmetric or even function, the direct product, $(\psi_{0(v)} \times \psi_{1(v)})$ gives back the symmetry of $\psi_{1(v)}$ and hence, the equation takes the form as:

$$M = \langle \psi_{0(e)} | \mu_e | \psi_{1(e)} \rangle \langle \psi_{1(v)} \rangle$$

Therefore, the nature of the whole integral ultimately depends on the symmetry of the final vibrational mode, $\psi_{1(v)}$ or, $(\psi_v)_f$ and it must be of same symmetry as the triple product of $[\psi_{0(e)} \times \mu_e \times \psi_{1(e)}]$ so that overall symmetry becomes totally symmetric or even in nature. This is bacause of the fact that two similar types of symmetries produce totally symmetric representation.

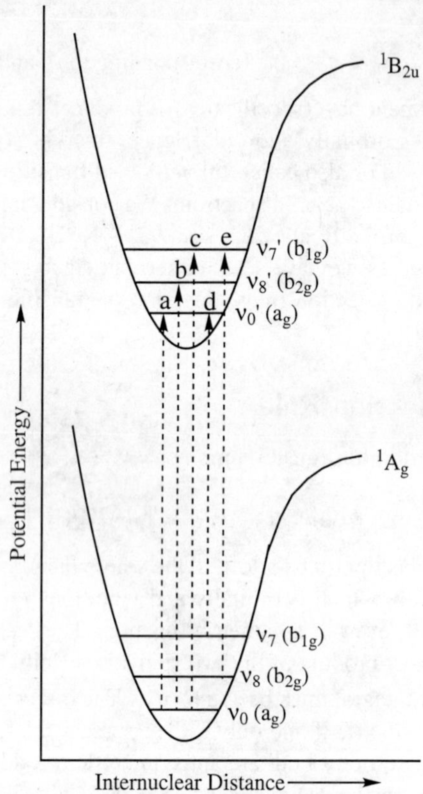

Figure 9.1: Few selective vibronic transitions of *p*-difuorobenzene.
(*Ref.* Symmetry and spectroscopy by D. C. Harris and M. D. Bertolucci, Dover publication, 1989)

Let us consider some selected vibronic transitions of a molecule of D_{2h} symmetry, *p*-difuorobenzene (shown in Figure 9.1). Two vibrational modes (ν_7 and ν_8) and their corresponding energy states in ground and excited electronic state of the molecule have been considered. Each of the total thirty normal mode of vibrations of *p*-difuorobenzene would require a different potential well to describe its all possible energy levels. Here for convenience two vibrational modes with different transitions in a single two-dimensional Morse-like diagram have been shown. The vibrational ground state of the ground electronic state (lower potential well) is called ν_0 and represents the state in which all the vibrations (here, only ν_7 and ν_8 are shown) occurs with their zero-point energies as:

$$\bar{\nu_0} = \sum_{i=1}^{3N-6} (1/2)\bar{\omega}_i$$

The same description is applied for the excited electronic state as well. The $0-0$ transition has been labeled as 'a', whereas 'b' ,'c' and 'd' are designated as 8^1_0 (ν_8 mode is excited from $v=0$ to $v'=1$), 7^1_0 (ν_7 mode is excited from $v=0$ to $v'=1$) and 7^0_1 (ν_7 mode is excited from $v=1$ to $v'=0$, also called *hot band*) respectively. The last mode of vibration originates from an excited vibrational state in the ground electronic state. Transition 'e' is a combination of ν_7 and ν_8 modes of vibration and is represented by $8^0_1 7^1_0$.

9.1.4.1 0^0_0 Transition:

Let us check the transition activity of a spin allowed pure electronic transition (0-0 transition) of *p*-difluorobenzene, $^1A_g \rightarrow {}^1B_{2u}$. It occurs from ground vibrational state ($v=0$) of ground electronic state to ground vibrational state ($v'=0$) of excited electronic state. Since, there is no contribution of any vibrational degrees of freedom, hence the name of the vibration is given as "0". Therefore, in 0^0_0 notation, "0" indicates the name or number of vibrational mode, subscript zero corresponds to the ground vibrational level of ground electronic state (from where transition originates) and superscript zero corresponds to the vibrational level of excited electronic state where the transition terminates. The same notation is applied for other vibronic transitions. The direct products due to transition moments are obtained as:

$$B_{2u} \times B_{3u} \times A_g = B_{1g} \text{ (corresponding to } \mu_x)$$

$$B_{2u} \times B_{2u} \times A_g = A_g \text{ (corresponding to } \mu_y)$$

$$B_{2u} \times B_{1u} \times A_g = B_{3g} \text{ (corresponding to } \mu_z)$$

Hence, this transition becomes allowed due to the presence of the y-component of dipole component and hence, is called to be y-polarised transition.

Now instead of pure electronic transition, if we revisit the same transition considering as vibronic transition we find the same result since, both $\psi_{v(0)}$ and $\psi_{v(1)}$ transform as totally symmetric, a_g (Since both correspond to vibrational ground state in two different electronic states), which is evident from the direct products shown below. Here we have to consider both electronic and vibrational wave functions in both the initial and final states.

$$(B_{2u} \times A_g) \times B_{3u} \times (A_g \times A_g) = B_{1g} \text{ (corresponding to } \mu_x)$$

$$(B_{2u} \times A_g) \times B_{2u} \times (A_g \times A_g) = A_g \text{ (corresponding to } \mu_y)$$

$$(B_{2u} \times A_g) \times B_{1u} \times (A_g \times A_g) = B_{3g} \text{ (corresponding to } \mu_z)$$

9.1.4.2 8^1_0 Transition:

It is a transition of ν_8 mode of vibration from ground electronic (A_g symmetry) as well as ground vibrational state (A_g symmetry) to first excited electronic state (B_{2u} symmetry) as well as to excited vibrational state of B_{2g} symmetry. Hence, considering symmetries of both electronic as well as vibrational wave functions, the direct products corresponding to three transition moments of three components of dipole moments are:

$$(B_{2u} \times B_{2g}) \times B_{3u} \times (A_g \times A_g) = B_{3g} \text{ (corresponding to } \mu_x)$$

$$(B_{2u} \times B_{2g}) \times B_{2u} \times (A_g \times A_g) = B_{2g} \text{ (corresponding to } \mu_y)$$

$$(B_{2u} \times B_{2g}) \times B_{1u} \times (A_g \times A_g) = B_{1g} \text{ (corresponding to } \mu_z)$$

There is no component corresponding to totally symmetry, A_g in the result of direct products. Hence, this transition is expected to be forbidden.

9.1.4.3 7_0^1 Transition:

Similarly, considering symmetries of electronic as well as vibrational wave functions for ν_7 mode , the direct products are:

$$(B_{2u} \times B_{1g}) \times B_{3u} \times (A_g \times A_g) = A_g \text{ (corresponding to } \mu_x)$$

$$(B_{2u} \times B_{1g}) \times B_{2u} \times (A_g \times A_g) = B_{1g} \text{ (corresponding to } \mu_y)$$

$$(B_{2u} \times B_{1g}) \times B_{1u} \times (A_g \times A_g) = B_{2g} \text{ (corresponding to } \mu_z)$$

Hence, this transition is spin as well as orbital allowed and also x-polarised.

9.1.4.4 Hot Band Transition, 7_1^0:

$$(B_{2u} \times A_g) \times B_{3u} \times (A_g \times B_{1g}) = A_g \text{ (corresponding to } \mu_x)$$

$$(B_{2u} \times A_g) \times B_{2u} \times (A_g \times B_{1g}) = B_{1g} \text{ (corresponding to } \mu_y)$$

$$(B_{2u} \times A_g) \times B_{1u} \times (A_g \times B_{1g}) = B_{2g} \text{ (corresponding to } \mu_z)$$

Hence, this is an allowed transition and x-polarised too. However, it must be kept in mind that since it is a hot band, such higher vibrational level may hardly be populated in the electronic ground state. Therefore, although it is a fully allowed transition but the intensity will be very low.

9.1.4.5 Combination Transition, $8_1^0 7_0^1$:

Considering the corresponding symmetries of electronic and vibrational wave functions, the three direct products would be:

$$(B_{2u} \times B_{1g}) \times B_{3u} \times (A_g \times B_{2g}) = B_{2g} \text{ (corresponding to } \mu_x)$$

$$(B_{2u} \times B_{1g}) \times B_{2u} \times (A_g \times B_{2g}) = B_{1u} \text{ (corresponding to } \mu_y)$$

$$(B_{2u} \times B_{1g}) \times B_{1u} \times (A_g \times B_{2g}) = A_g \text{ (corresponding to } \mu_z)$$

Hence, this combination transition is found to be allowed and z-polarized as well.

9.2 Few Examples

Now, taking a few specific examples, let us see how the nature of electronic transition alters with the nature of orbital symmetry.

9.2.1 Formaldehyde

There are total three σ bonding molecular orbitals (two σ_{C-H} and one $\sigma_{\mathrm{C-O}}$) and three corresponding σ^* anti-bonding orbitals in formaldehyde molecule. There is also one π and one π^* anti-bonding orbital in C−O bond and one pair of orbitals on the oxygen atom as well that accommodate the non-bonding electrons (n-orbital). The symmetries of those molecular orbitals can be derived by performing different symmetry operations corresponding to the C_{2v} point group and observing how the orbitals (basis) gets transformed. The symmetries along with their characters corresponding to each symmetry operations have been shown in Table 9.1.

Table 9.1: Effect of symmetry operations on different kind of molecular orbitals of formaldehyde molecule.

Basis (Orbitals)	E	$C_2(z)$	σ_{xz}	σ_{yz}	*Symmetry*
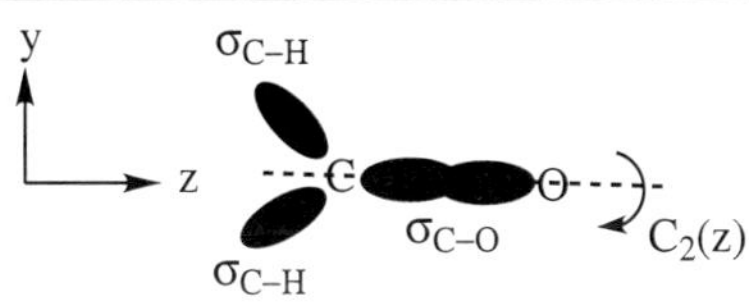	1	1	1	1	A_1
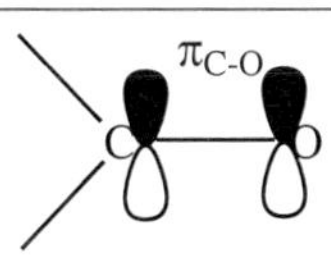	1	−1	−1	1	B_2
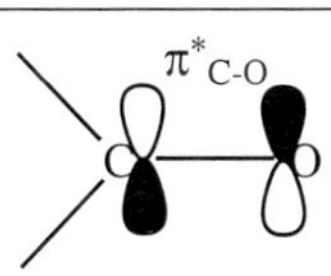	1	−1	−1	1	B_2
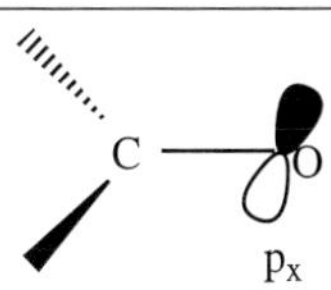	1	−1	1	−1	B_1
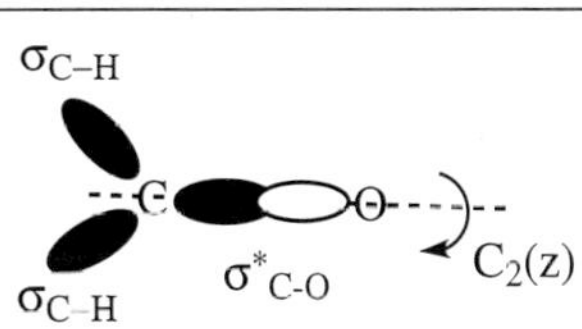	1	1	1	1	A_1

It is known that the ground state electronic configuration of formaldehyde is $\sigma^2_{(C-O)}\pi^2_{(C-O)}\,n^2$. The symmetry of the ground state can be obtained from the overall direct product of the symmetries of all the orbitals involved, *i.e.*,

$$(A_1)^2 \times (B_2)^2 \times (B_1)^2 = A_1$$

All doubly occupied orbital basically results totally symmetric representation.

Now if an electronic transition takes place from π to π^* orbital, the electronic configuration becomes:

$$\sigma^2_{(C-O)}\ \pi^1_{(C-O)}\ n^2\pi^{*1}_{(C-O)}$$

Hence, the symmetry of the $\pi\pi^*$ excited state is given by the following direct product:

$$(A_1)^2 \times (B_2) \times (B_1)^2 \times (B_2) = A_1$$

Similarly, for $n - \pi^*$ transition, the electronic configuration becomes:

$$\sigma^2_{(C-O)}\ \pi^2_{(C-O)}\ n^1\pi^{*1}_{(C-O)}$$

The corresponding symmetry of the $n\pi^*$ excited state is:

$$(A_1)^2 \times (B_2)^2 \times (B_1) \times (B_2) = A_2$$

For $n - \sigma^*$ transition, the electronic configuration is:

$$\sigma^2_{(C-O)}\ \pi^2_{(C-O)}\ n^1\ \sigma^{*1}_{(C-O)}$$

And corresponding symmetry of $n - \sigma^*$ excited state is

$$(A_1)^2 \times (B_2)^2 \times (B_1) \times (A_1) = B_1$$

Selection Rule

It is known that the intensity of electronic transition is proportional to the integral $\langle\psi_g|\mu||\psi_e\rangle$. Now the symmetry of the ground and excited states are just determined. The symmetry of the three dipole moment components (μ_x, μ_y and μ_z) are B_1, B_2 and A_1 respectively as evident from the character table. For a transition from ground state to each of the excited states, there will be three integrals for three components of μ. If at least one of the three integrals becomes totally symmetric (here it is A_1), then that transition will be allowed.

- For ground state to $\pi\pi^*$ excited state, the three integrals are:

$$\langle A_1 \times B_1 \times A_1\rangle = B_1$$

$$\langle A_1 \times B_2 \times A_1\rangle = B_2$$

$$\langle A_1 \times A_1 \times A_1\rangle = A_1$$

  Since, the last integral results A_1, hence this transition is symmetrically allowed and z-polarized.

- For ground state to $n\pi^*$ excited state, the three integrals are:

$$\langle A_1 \times B_1 \times A_2 \rangle = B_2$$

$$\langle A_1 \times B_2 \times A_2 \rangle = B_1$$

$$\langle A_1 \times A_1 \times A_2 \rangle = A_2$$

Since, neither of the three integrals result A_1 symmetry, hence, symmetrically forbidden.

- For ground state to $n\sigma^*$ excited state, the three integrals are:

$$\langle A_1 \times B_1 \times B_1 \rangle = A_1$$

$$\langle A_1 \times B_2 \times B_1 \rangle = A_2$$

$$\langle A_1 \times A_1 \times B_1 \rangle = B_1$$

Since, the first integral results A_1 symmetry,hence, this transition is symmetrically allowed and x-polarized.

9.2.2 Water

It is known that the ground state of water molecule (C_{2v}) is: $(1a_1)^2(1b_2)^2(2a_1)^2(1b_1)^2$. The nature of ground state is totally symmetric, A_1. The two excited electronic states originated due to different transitions of the molecule are as follows:

- **$b_1 \rightarrow a_1^*$ transition:**
  The excited state electronic configuration: $(a_1)^2(b_2)^2(a_1)^2(b_1)(a_1^*)$
  Symmetry: $(b_1)(a_1) = B_1$
  The transition dipole moment integrals:
  For x-component: $A_1 \times B_1 \times B_1 = A_1$
  For y-component: $A_1 \times B_2 \times B_1 = A_2$
  For z-component: $A_1 \times A_1 \times B_1 = B_1$
  Hence, we see that only x-component results totally symmetric representation. Therefore, it can be concluded that this transition is allowed and x-polarised.

- **$b_1 \rightarrow b_2^*$ transition**
  The excited state electronic configuration: $(a_1)^2(b_2)^2(a_1)^2(b_1)(b_2^*)$
  Symmetry:$(b_1)(b_2) = A_2$
  The transition dipole moment integrals:
  For x-component: $A_1 \times B_1 \times A_2 = B_2$
  For y-component: $A_1 \times B_2 \times A_2 = B_1$
  For z-component: $A_1 \times A_1 \times A_2 = A_2$
  Here, we see that the totally symmetric representation is absent in all the integrals. Therefore, it is concluded that this transition is forbidden.

9.2.3 Butadiene

The ground state electronic configuration of butadiene (C_{2h}) is $(a_u)^2(b_g)^2$ with symmetry of A_g.

The four excited electronic states of the molecule are as follows:

$(a_u)^2(b_g)(a_u^*)$; Symmetry: $(b_g)\times(a_u) = B_u$

$(a_u)^2(b_g)(b_g^*)$; Symmetry: $(b_g)\times(b_g) = A_g$

$(a_u)(b_g^2)(a_u^*)$; Symmetry: $(a_u)\times(a_u) = A_g$

$(a_u)^2(b_g^2)(b_g^*)$; Symmetry: $(a_u)\times(b_g) = B_u$

Out of these four probable transitions, $b_g \rightarrow a_u^*$ and $a_u \rightarrow b_g^*$ transitions have alternate but similar initial and final state symmetries. Similarly, $b_g \rightarrow b_g^*$ and $a_u \rightarrow a_u^*$ transitions have alternate but similar initial and final state symmetries. As evident from the character table, three dipole moment components corresponding to x, y and z, have the symmetries as B_u, B_u and A_u.

Let us discuss the spectroscopic activities of these transitions.

- $b_g \rightarrow a_u^*$ and $a_u \rightarrow b_g^*$ **transitions**:
  For x-component: $A_g \times B_u \times B_u = A_g$
  For y-component: $A_g \times B_u \times a_g = A_g$
  For z-component: $A_g \times A_u \times B_u = B_g$
  Since, we see that x-component and y-component result the totally symmetric representation, A_g Hence, both the transitions are allowed and x and y-polarized as well.

- $b_g \rightarrow b_g^*$ and $a_u \rightarrow a_u^*$ **transitions**:
  For x-component: $A_g \times B_u \times A_g = B_u$
  For y-component: $A_g \times B_u \times A_g = B_u$
  For z-component: $A_g \times A_u \times A_g = A_u$
  Since, none of the components results the totally symmetric representation, both the transitions are expected to be forbidden. Also careful inspection of the results shows that $g \leftrightarrow u$ transitions are allowed and $g \leftrightarrow g$, $u \leftrightarrow u$ transitions are forbidden.

9.2.4 Benzene

There are six π-MOs for benzene molecule. The construction of six π-MO's for benzene molecule (point group D_{6h}) has been discussed using projection operation technique in chapter 11. The six π-MO's are as follows:

$$\psi_1(a_{2u}) = \left(\frac{1}{\sqrt{6}}\right)(\phi_1+\phi_2+\phi_3+\phi_4+\phi_5+\phi_6)$$

$$\psi_2(e_{1g}) = \left(\frac{1}{\sqrt{12}}\right)(2\phi_1+\phi_2-\phi_3-2\phi_4-\phi_5-\phi_6)$$

$$\psi_3(e_{1g}) = \left(\frac{1}{2}\right)(\phi_2+\phi_3-\phi_5-\phi_6)$$

$$\psi_4(e_{2u}) = \left(\frac{1}{\sqrt{12}}\right)(2\phi_1 - \phi_2 - \phi_3 + 2\phi_4 - \phi_5 - \phi_6)$$
$$\psi_5(e_{2u}) = \left(\frac{1}{2}\right)(\phi_2 - \phi_3 + \phi_5 - \phi_6)$$
$$\psi_6(b_{2g}) = \left(\frac{1}{\sqrt{6}}\right)(\phi_1 - \phi_2 + \phi_3 - \phi_4 + \phi_5 - \phi_6)$$

Under the Huckel approximations, the energies of the above MOs can be worked out by using the expression $\langle\psi|H|\psi^*\rangle$. Their energies and corresponding symmetries are shown in Figure 9.2.

$$E_1 = \alpha + 2\beta,\ E_2 = E_3 = \alpha + \beta,\ E_4 = E_5 = \alpha - \beta,\ E_6 = \alpha - 2\beta$$

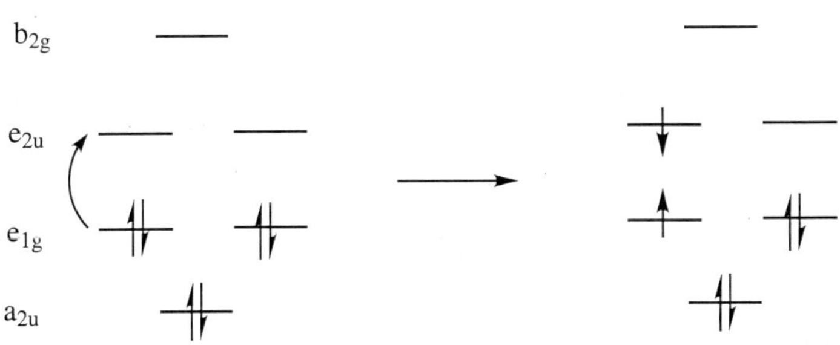

Ground state (Symmetry: A_{1g}) **Excited state** (Symmetry: B_{1u} +B_{2u} +E_{1u})

Figure 9.2: Electronic configuration of benzene in ground as well as excited states with E_1 as lowest energy and E_6 as highest energy.

Since, benzene molecule contains six π-electrons, the ground state symmetry is a_{1g} with a configuration of

$$(A_{2u})^2(e_{1g})^4$$

The electronic spectrum of benzene involves the transitions from the e_{1g} level to available e_{2u} and b_{2g} levels. However, the transition $e_{1g} \rightarrow b_{2g}$ involves transition between same parity, $g \leftrightarrow g$ and thus is forbidden. The promotion of one electron from e_{1g} to e_{2u} will lead to the excited state, $(a_{2u})^2(e_{1g})^3(e_{2u})^1$. To get the symmetry of this excited state, we require the symmetries of $(e_{1g})^3$ and $(e_{2u})^1$. The direct product, $(e_{1g} \times e_{1g} \times e_{1g})$ does not give the correct result for the orbital symmetry. From quantum mechanics, it is known that the symmetry properties of $(e_{1g})^3$ state is identical with that of $(e_{1g})^1$. In general, if a shell has a total capacity of n electron (here in this orbital,the total capacity is 4), then the symmetry properties of an electronic configuration containing m electrons (here the number is 3) are exactly identical to that containing *(n-m)* electrons (here 4 – 3 = 1), known as *hole formalism*. Hence, the symmetry of the above excited state can be obtained by direct product of e_{1g} and e_{2u} which gives us the following reducible representation:

Table 9.2: Reducible representation.

D_{6h}	E	$2C_6$	$2C_3$	C_2	$3C'_2$	$3C''_2$	i	$2S_3$	$2S_6$	σ_h	$3\sigma_d$	$3\sigma_v$
$\Gamma(e_{1g} \times e_{2u})$	4	–1	1	–4	0	0	–4	1	–1	4	0	0

After reduction of the resultant reducible representation it is found,

$$\Gamma = B_{1u} + B_{2u} + E_{1u}.$$

Hence, the nature of excited state is not of a single symmetry, rather combination of three symmetries. Therefore, for determination of transition dipole moment integrals, we have to consider each of the symmetries simultaneously for excited state for each dipole moment component.

If we consider the μ_z component, following the character table, the symmetry is known as A_{2u} type. The ground state is as usual, totally symmetric, A_{1g} type. We need to take the third component of the integral, *i.e.*, the symmetry of the excited state as B_{1u}, B_{2u} or E_{1u} simultaneously. Hence, the three integrals corresponding to μ_z component would be:

$$A_{1g} \times A_{2u} \times B_{1u} = B_{2g}$$

$$A_{1g} \times A_{2u} \times B_{2u} = B_{1g}$$

$$A_{1g} \times A_{2u} \times E_{1u} = E_{1g}$$

Since, the resultants components of excited state, B_{2g} or B_{1g} or E_{1g} does not contain or involve any totally symmetric representation, hence, along the z-component the transition is said to be forbidden.

Similarly, if we consider the x and y components (from character table, x and y both dipole moment components have E_{1u} symmetry), we get

$$A_{1g} \times E_{1u} \times B_{1u} = E_{2u}$$

$$A_{1g} \times E_{1u} \times B_{2u} = E_{2g}$$

$$\begin{aligned} A_{1g} \times E_{1u} \times E_{1u} &= A_{1g} \times (A_{1g} + A_{2g} + E_{2g}) \\ &= A_{1g} \times A_{1g} + A_{1g} \times A_{2g} + A_{1g} \times E_{2g} \\ &= A_{1g} + A_{2g} + E_{2g} \end{aligned}$$

Since, the direct product, $A_{1g} \times E_{1u} \times E_{1u}$ contains the totally symmetric representation, A_{1g} as one of the components, the transition, $A_{1g} \rightarrow E_{1u}$ is said to be symmetry allowed.

9.3 *d-d* Transition

This is one of the most important mechanism for the origin of colour in most of the transition metal complexes. Here in this topic, how the symmetry plays crucial role in explaining the mechanism, would be discussed. As already, this is very much known to us that the transition moment integral for two electronic states are as follows:

$$\langle \Psi_0 \begin{pmatrix} \mu_x \\ \mu_y \\ \mu_z \end{pmatrix} \Psi_1 \rangle$$

In case of *d-d* transition, the wave functions of initial ground state and final excited state are symmetric or even function, which is also evident from their shapes (All the *d*-orbitals possess inversion centre). Hence the nature of parity is of *g* type for *d*-orbitals.

The nature of μ_x, μ_y and μ_z depends on the symmetry of a molecule or the nature of a point group. For example, in octahedral point group, μ_x, μ_y and μ_z corresponds to T_{1u} symmetry, *i.e.*, *u* type, found from character table. Therefore, the transition moment integral becomes odd type or *u* in nature.

$$\text{Therefore, } M = \langle g|u|g\rangle = u$$

Hence, the integral would definitely produce a zero value, which consequently depicts that *d-d* transition is forbidden in terms its parity. Similarly, *s-s* and *p-p* transitions are also forbidden from parity ground.

For s-s- transition, $M = \langle g|u|g\rangle = u$

and, p-p transition, $M = \langle u|u|u\rangle = u$

These transitions are also forbidden from *Laporte* selection rule as it says that there must be a change in angular momentum or, $\Delta\,l = \pm 1$.

In *s-s*, *p-p* or *d-d* transitions, there are no change of angular momentum, where as, *s-p*, *p-d* transitions are accompanied by change of angular momentum and hence, become allowed. In these cases, both parity as well as angular momentum changes.

Now the question is in spite of the forbidden nature of *d-d* transitions, how it becomes allowed? In most of the octahedral transition metal complexes shows significant colours suggesting the *d-d* transition. Since, except very few cases, electronic transitions are not independent of vibrational motions of a molecule. Here the vibronic coupling plays an important role to make *d-d* transition allowed. Vibrational movements remove the inversion symmetry of a molecule, which is basically prerequisite criteria for parity selection rule.

Already it has been discussed that in transition moment integral,

$$M = \langle(\psi_e)_i\, \mu_e (\psi_e)_f\rangle(\psi_v)_f$$

the nature of the integral corresponding to electronic part bears *u* character. Hence, M would be non-zero only when the vibrational mode, $(\psi_v)_f$ is also of similar nature, *i.e.*, *u* type. An octahedral molecule of AB_6 type, is comprised of total 15 vibrational modes. Out of these 15 modes, only the vibrations which are of antisymmetric in nature (u) would be expected to make *d-d* transition allowed.

Let us take an example of d^1 system where $2T_{2g} \rightarrow 2E_g$ transition is considered in an octahedral environment.

The transition moment integral,

$$M = \langle(\psi_e)_i\, \mu_e (\psi_e)_f\rangle(\psi_v)_f$$

$$= (T_{2g} \times T_{1u} \times E_g) \times (\psi_v)_f$$

The triple product, $(T_{2g} \times T_{1u} \times E_g)$ gives a reducible representation and after reduction, it results $A_{1u} + A_{2u} + 2E_u + 2T_{1u} + 2T_{2u}$

Already it is known that $\Gamma_{vib} = A_{1g} + E_g + 2T_{1u} + T_{2g} + T_{2u}$

$$\text{Hence, } M = (A_{1u} + A_{2u} + 2E_u + \underline{2T_{1u}} + \underline{2T_{2u}}) \times (A_{1g} + E_g + \underline{2T_{1u}} + T_{2g} + \underline{T_{2u}})$$

Here, we do not need to carry out the whole direct product, but since in both the terms similar symmetries are present (underlined symmetries), hence, the product is going to contain definitely a totally symmetric term. Therefore, $M \neq 0$ *i.e.*, transition is allowed.

Although, the parity rule was making this transition forbidden, but due to the presence of vibration, not only it lifts the parity rule, but also due to mixing of p- and d orbitals, the pure d-character is lost and hence during transition, $\Delta l \neq 0$.

In **tetrahedral geometry** for d-d transition, there is no inversion centre and hence no question of parity rule, but still due to same angular momemntum, $\Delta l = 0$, $d - d$ transition is predicted to be forbidden. However, p and d-orbital mixing is symmetry allowed under T_d symmetry (since, p orbitals have same t_2 symmetry as d_{xy}, d_{xz} and d_{yz}).

Here one point must be noted that as a result of the parity restriction in O_h, the colour of T_d complexes is generally more intense than O_h complexes.

EXERCISES

9.1 Water molecule has the ground state electronic configuration as $(1a_1)^2(1b_2)^2(2a_1)^2(1b_1)^2$. Find whether the lowest energy transition is allowed or forbidden. Also find the allowedness of other probable transitions.

9.2 Ground state electronic configuration of *trans*-butadiene is $(a_u)^2(b_g)^2$. Comment on the allowedness of transitions from HOMO to other lower energy antibonding orbitals.

9.3 What is *hole formalism*? What is its advantage? Show one example of *hole formalism*?

9.4 Why generally tetrahedral complexes have more intense color than corresponding octahedral complexes?

9.5 Show that the symmetry of excited state of benzene do not correspond to a pure irreducible representation, rather a combination of several symmetries. Also find the allowedness of transition.

Chapter 10

Applications to Chemical Bonding

10.1 Introduction

As we know that except the simplest diatomic molecule, exact solution of a large molecular wave function is not feasible. Therefore, so far it is a conventional technique to construct empirical wave functions by combining wave functions of various atoms of the molecule. This method of combining atomic orbitals is widely known as Linear Combination of Atomic Orbitals (LCAO) method. If we take a general case of diatomic molecule, AB, then the two possible combinations which form molecular orbitals are:

$$\Psi_1 = a\psi_A + b\psi_B$$

$$\Psi_2 = a\psi_A - b\psi_B$$

Here, it is seen that in one case, in-phase combination and in another case out-of-phase combination is observed. The coefficients a and b are some constants which reflect the relative contributions of the two participating atomic orbitals (AOs) in the linear combinations to construct the two molecular orbitals (MOs). At this point, we should keep in mind that for efficient construction of MOs, the three basic necessary conditions must be fulfilled:

- *The participating AOs must be of similar energy.*
- *There must be significant spatial overlap of AOs.*
- *They must possess similar symmetry with respect to internuclear axis.*

It is also known that hybridization is not an essential feature of the molecular orbital approach. The construction of MO for molecules like AB_n can be treated by matching the symmetries of orbitals on the central A atom with those of mathematical combinations of orbitals on the peripheral B atoms which are also called *pendant atoms*. These new orbitals, thus created from pendant atoms, are called *ligand group orbitals*

(LGO). The mathematical combinations of pendant atom AOs are called *symmetry adapted linear combinations* or SALCs. Very briefly, the two approaches of molecular orbital formation can be summarized as follows: In hybridization approach the main focus is on the central atom. According to the electronic configuration of the central atom, the hybridization takes place and accordingly it spreads its hybrid orbitals in proper direction in space and then the peripheral atoms are connected to those hybrid orbitals to give the proper final shape of the molecule. Whereas, in SALC approach, the peripheral atoms or pendant atoms are important, where these atoms form orbitals of various combinations to give some symmetry type (SALC) and then those newly formed group of orbitals (LGOs) which are directed to the central atom, are combined according to the matching of symmetry with atomic orbitals of the central atom again by linear combination method.

When SALCs are applied the resulting LCAO-MO gets the form as,

$$\Psi_{MO} = a\psi_{AO}(A) \pm b\psi_{SALC}(nB)$$

Where, SALCs have the general form as,

$$\psi_{SALC} = c_1\psi_1 \pm c_2\psi_2 \pm c_3\psi_3 \pm \ldots\ldots c_n\psi_n$$

At this point, we can write down the whole process of MO formation by step wise as follows:

- Potentially bonding orbitals of peripheral atoms are taken as basis.
- From these basis orbitals, the LGOs are formed through linear combinations and from transformation properties of those orbitals, reducible representation is derived.
- After reduction to its component irreducible representations, their nature of symmetry is easily found taking the help of corresponding character table.
- The symmetry nature of participating valence orbitals of central atom is obtained with the help of character table.
- Through symmetry matching, these two type of orbitals are combined to form SALC-MO.

After constructing MOs, we need to keep few general points in mind when these MO energy levels are organized.

1. Bonding MOs formed by in-phase combination of orbitals always possess lower energy than corresponding antibonding MO formed through out-of-phase combination of orbitals.
2. Nonbonding MOs remains unaltered in energy and occupy the position between bonding and antibonding energy levels.
3. Since, π-interaction is always less efficient than σ-interaction, hence π-bonding MOs tend to have higher energies than σ-bonding MOs formed from similar AOs. Similarly, π-antibonding MOs tend to be less antibonding in character and have less energies than σ-antibonding MOs formed from similar AOs.
4. MO energies increases with the increase of number of nodes. MO with lowest number of nodes correspond to lowest energy level.

5. Among σ-bonding MOs, lowest energy level is generally considered as the totally symmetric in nature.

Before proceeding further into LGO approach, we can see some applications of symmetry aspects in finding the appropriate atomic orbitals of central atom during hybridization process in chemical bonding.

10.2 Symmetry of Hybridization of Different Molecules

10.2.1 H_2O (C_{2v} Point Group)

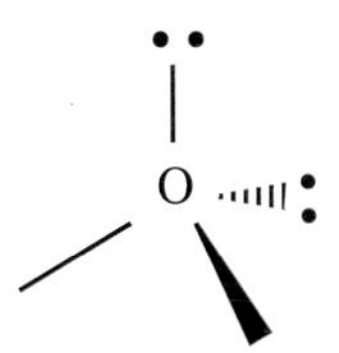

Water molecule has the distorted tetrahedron geometry as it has two bonded and two lone pairs of electrons. Let us take the four hybrid orbitals of O-atom as bases and are represented by $\phi_1, \phi_2, \phi_3, \phi_4$. The reducible representation will be as follows:

C_{2v}	E	C_2	σ_{xz}	σ_{yz}
Γ_{Red}	4	0	2	2

After reduction we get,

$$\Gamma_{Red} = 2A_1 + B_1 + B_2$$

From the character table, two A_1 symmetries correspond to s and p_z orbitals, B_1 correspond to p_x and B_2 correspond to p_y orbital. Thus, we see that the hybridization involved is sp^3 in water molecule.

10.2.2 NH_3 (C_{3v} Point Group)

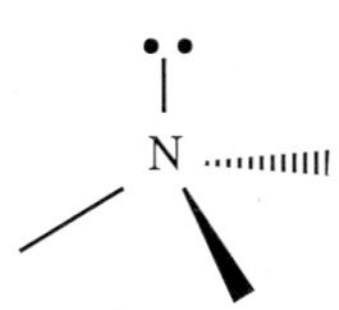

In ammonia, there are three bonded and one lone-pair of electrons. Let the four hybrid orbitals ϕ_1, ϕ_2, ϕ_3 and ϕ_4 are taken as basis and try to develop the reducible representation. On application of E-operation, all the four bases remains unchanged, C_3 and C_3^2 operations leave only the basis corresponding to lone pair orbital unaltered and each σ_v keeps one lone pair orbital and one bonded orbital basis unaltered. Hence, the table corresponding to reducible representation looks as shown below.

C_{3v}	E	$2C_3$	$3\sigma_v$
Γ_{Red}	4	1	2

After reduction we get,

$$\Gamma_{Red} = 2A_1 + E$$

The orbitals corresponding to the symmetry A_1 and E combine to give hybrid orbitals. From the character table of C_{3v}, it can be seen that the two A_1 symmetries represent the totally symmetric s and p_z orbitals and one degenerate E symmetry represent p_x and p_y orbitals. Thus, it can be concluded that the hybridization involved in this molecule is sp^3.

10.2.3 SF$_6$ (O_h Point Group)

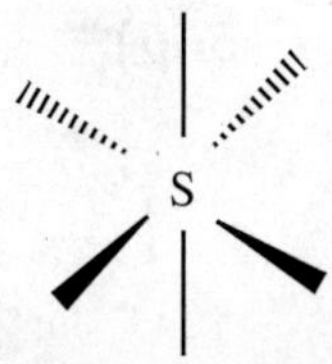

It belongs to octahedral(O_h) point group. The various classes of symmetry elements are E, C_3, C_2, C_4, $C_4^2 = C_2$, i, S_4, S_6, σ_h, σ_d. Let the six hybrid orbitals $\phi_1, \phi_2, \phi_3, \phi_4, \phi_5$ and ϕ_6 are taken as basis and try to develop the reducible representation by simply calculating the number of shifted and unshifted basis.

O_h	E	C_3	C_2	C_4	$C_4^2 = C_2$	i	S_4	S_6	σ_h	σ_d
Γ_{Red}	6	0	0	2	2	0	0	0	4	0

After reduction this reducible representation, we get,

$$\Gamma_{Red} = A_{1g} + E_g$$

From character table it is evident that totally symmetric, A_{1g} corresponds to the most symmetric s orbital, doubly degenerate, E_g correspond to two d orbitals namely $d_{x^2-y^2}$ and d_{z^2} and T_{1u} corroborates to three triply degenerate p_x, p_y and p_z orbitals. Thus the possible hybridization will be consists of one s orbital, three p orbitals and two d orbitals and hence, can easily be predicted as sp^3d^2.

10.2.4 CH$_4$ (T_d Point Group)

Here we assume the tetrahedral structure of methane molecule and will try to find the nature of atomic orbitals involved in hybridization.

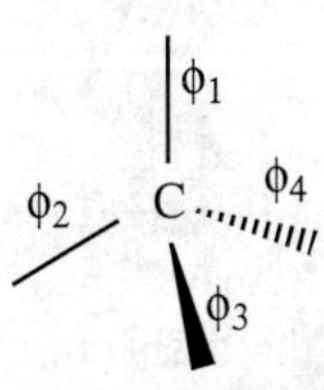

Let us take the four hybrid orbitals of methane as bases and are represented as ϕ_1, ϕ_2, ϕ_3 and ϕ_4 as shown in the figure. Now we will try to find the character representation (or reducible representation) of each symmetry operation on those hybrid orbitals.

T_d	E	$8C_3$	$3C_2$	$6S_4$	$3\sigma_d$
Γ_{hybrid}	4	1	0	0	2

Taking the help of T_d character table, we found the corresponding IR components as:

$$\Gamma_{hybrid} = A_1 + T_2$$

A_1 is the totally symmetric representation and it corresponds to highly symmetric s orbital which is also evident from the presence of $x^2+y^2+z^2$ in the character table. Similarly, the presence of x, y and z corresponds to p_x, p_y and p_z orbitals in T_2 symmetry IR. Thus, we can say that the hybrid orbitals are formed by mixing of one s-orbital and three p-orbitals resulting in sp^3 hybridization.

In T_2 symmetry along with x, y and z there are binary terms like xy, yz and xz too which corresponds to d_{xy}, d_{yz}, d_{xz} orbitals. Hence, alternatively there is a possibility of sd^3 hybridization as well. According to symmetry considerations, both sets sp^3 and sd^3 are equally possible to constitute hybrid orbitals. Hence, from symmetry consideration we can not proceed further. Now we can proceed by using a linear combination of these two, *i.e.*,

$$\Psi_C = a\psi_{(sp^3)} + b\psi_{(sd^3)}$$

(Where, Ψ_C is hybrid orbital function for central C-atom.)

a and b can be calculated from quantum mechanical treatment and the relative magnitudes of these two depicts the relative contribution in formation of the overall hybrid orbital of C atom.

But in C atom, the excitation of electrons from the $2p$ level to $3d$ level involves very high energy (230 kcal) and hence this combination is energetically not favorable although symmetrically allowed. Quantum mechanics finds that the value of b is much less than a. Hence, the contribution due to sd^3 combination can be neglected very easily, *i.e.*

$$\Psi_C \approx \psi_{(sp^3)}$$

It may be noted that sp^3 hybridization energetically as well as symmetrically more favorable. However, for species like MnO_4^-, CrO_4^-, to constitute four equivalent hybrid orbitals sd^3 hybrid orbitals dominate.

10.2.5 BCl_3 (D_{3h})

Let us take the example of a trigonal planar molecule BCl_3 of point group, D_{3h} and construct the set of hybrid orbitals. Our job is to find which orbitals can be hybridized to produce a set of three trigonal planar σ-bonds. A set of three vectors (a_1, a_2, a_3) can be used to represent a set of trigonal planar bonds as shown in Figure 10.1. This set of three vectors is used as basis to generate reducible representation of D_{3h} point group. It is known that, depending upon the change of position/direction with respect to the symmetry operations of the point group gives the character of corresponding operation and is presented in the table below. Taking the help of character table, if

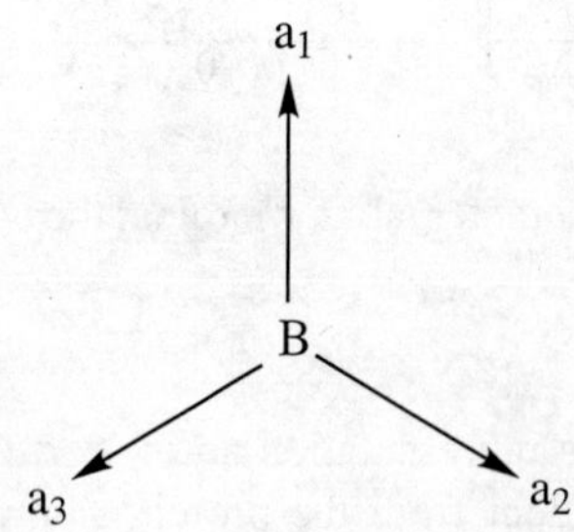

Figure 10.1: Three vectors as basis to generate reducible representation.

D_{3h}	E	$2C_3$	$3C_2$	σ_h	$2S_3$	$3\sigma_v$
Γ_1	3	0	1	3	0	1

we reduce the above representation to its constituent IRs, we get,

$$\Gamma_1 = A_1' + E'$$

If we now go through the character table again, we can see that A_1' includes either d_{z^2} or spherically symmetrical s orbital. E' includes p_x, p_y together and $d_{x^2-y^2}$, d_{xy} together. The two orbitals, p_x and p_y are degenerate as are $d_{x^2-y^2}$ and d_xy orbitals since they are bracketed together in the character table. It is most likely that for a first row element, Boron, the probable set of orbitals to form a trigonal hybrid would be s, p_x and p_y, (*i.e.*, one sp^2 set). The molecular plane is conventionally taken to be the xy plane whereas, z is taken as vertical.

In summary, first we chose a set of vectors representing the bonds as basis of our reducible representation. Then it was reduced to its components, $A_1' + E'$. After that, the results are interpreted to find the set of orbitals needed for hybridization.

The most crucial step in this process is the first one- the correct choice of basis. It must reflect the question or problem which we are trying to solve.

Now, let us see which orbitals would be suitable for π-bonding for a D_{3h} molecule. We know that a π-bonding has wave function whose sign differs in the two lobes (Figure 10.2a).

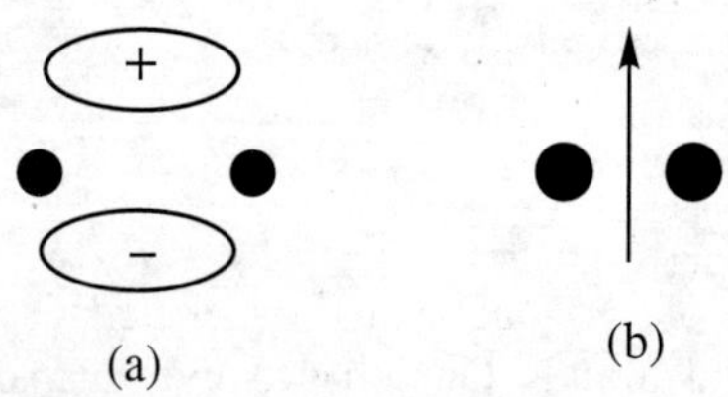

Figure 10.2: Representation of π-orbitals.

One arrow is drawn indicating the head towards the positive lobe, which represents the symmetry properties of the π-orbital (Fig. 10.2b). Remember that each pair of atoms could be linked by two π-bonds at right angles, and draw a suitable set of six arrows to act as a basis in π-bonds for a D_{3h} molecule of formula AB_3.

Two independent sets can be assumed which are potent candidate for π- bond formation; (a_1, a_2, a_3) is the out-of-plane set and (a_4, a_5, a_6) is the in-plane set as shown in Figure 10.3. Since the two sets would not be mixed by any of the group operations, hence these two sets can be considered separately.

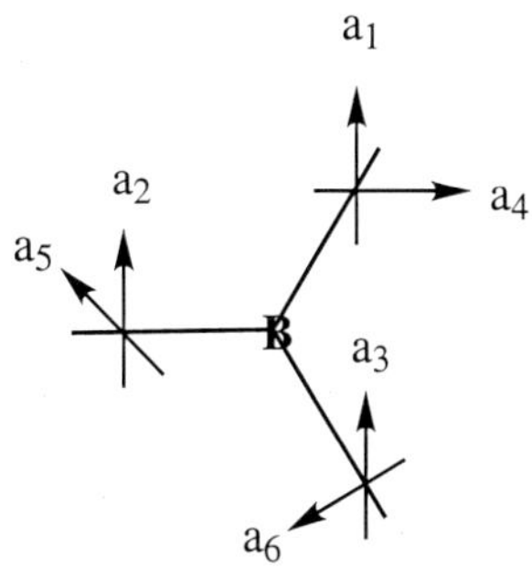

Figure 10.3: *In-plane* and *out-of-plane* vectors or basis.

Now, if we perform the group operations with respect to the set of *out-of-plane* (Γ_2) and set of *in-plane* orbitals (Γ_3) separately, the reducible representations generated as shown in the table below.

D_{3h}	E	$2C_3$	$3C_2$	σ_h	$2S_3$	$3\sigma_v$
Γ_2	3	0	–1	–3	0	1
Γ_3	3	0	–1	3	0	–1

After reduction of Γ_2 and Γ_3 we get:

$$\Gamma_2 \text{ (out-of-plane)} = A_2'' + E''$$

and

$$\Gamma_3 \text{ (in- plane)} = A_2' + E'$$

Now, we have to take the help of character table to decide which set of basis or orbitals are suitable for π-bonding between the two type of sets, *viz.*, *out-of-plane* or *in-plane* for this molecule. Character table depicts that, *out-of-plane* bases correspond to A_2'' IR which have the same symmetry nature as p_z orbital and correspond to degenerate E'' IR as well which have the same symmetry nature as d_{xy} and d_{yz} together. Similarly, *in-plane* bases correspond to degenerate E'' IR,which have the same symmetry nature as either p_x and p_y orbitals together or d_{xy} and $d_{x^2-y^2}$ orbitals together. It should be noted that A_2' IR does not correspond to any orbital, which is evident from character table.

For a first row element such as boron, there are no energetically available d orbitals. As already it is known that p_x and p_y orbitals are involved in σ-bonding in a molecule like BCl_3. Hence, we are left with only one orbital which is a genuine candidate for π-bond formation with respect to the whole molecular plane which is obviously p_z orbital. Therefore, this suggests that in Boron trihalides, there would be some π-bonding character. This would involve electrons being given from a filled chlorine p_z- orbital to the empty p_z orbital of boron. Pictorially all the probable σ and π-bonds are shown in Figure 10.4.

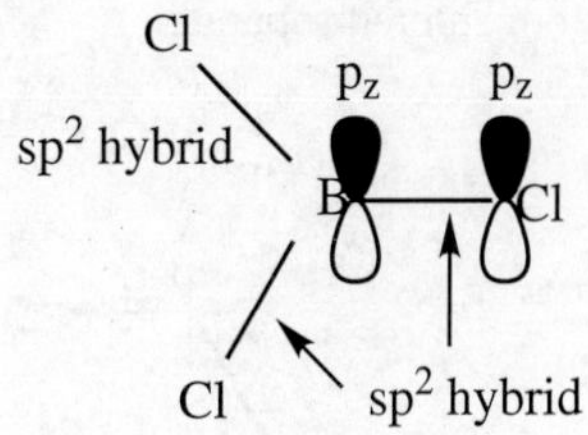

Figure 10.4: Type of bonds in BCl_3 molecule.

10.3 Ligand Group Orbitals (LGO)

We will conclude this chapter taking few examples where the concept of ligand group orbitals (LGO) have been used and SALC-MO have been constructed as well.

10.3.1 H_2O Molecule

For this we need to consider all the valence shell orbitals of central O-atom as well as two H-atoms. Let us consider the two 1*s* atomic orbitals of two H-atoms as ψ_1, ψ_2 and 2*s*, $2p_x$, $2p_y$, $2p_z$ orbitals of O-atom as ψ_3, ψ_4 ψ_5 and ψ_6 as shown in figure below.

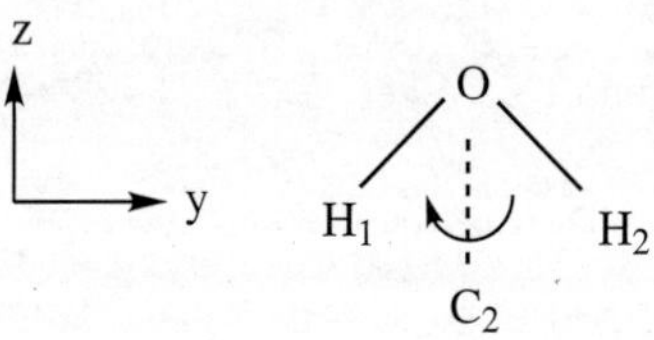

Whenever we are trying to construct MOs, always we need to keep in mind two criteria:

1. *Firstly, orbitals of only similar energy can interact efficiently. But it does not necessarily means that only orbitals of similar energy 1s-1s, or 2s-2s or 2p-2p orbitals will interact. It should be kept in mind that here in these cases the extent of interaction will be maximum and there will be a significant stabilization of the corresponding bonding MO and equally large extent of destabilization of the corresponding antibonding MO. As the difference in energy increases, the extent of interaction between AOs decreases. Also it is important to remember that if there is a difference in energy between the participating AOs then in the newly developed bonding MO the AO with lower energy will have more contribution and similarly in antibonding MO, the higher energy AO will have more contribution.*
2. *Secondly, there must be symmetry matching between the participating AOs. Again in a same way as the difference in symmetry increases, the extent of interaction decreases.*

Again during generation of MOs, we need to follow two basic steps:

1. *First we need to find out the way the AOs of the terminal atoms (here in water molecule, two H-atoms) combine which are basically known as ligand group orbitals (LGO).*
2. *In the next step we need to find out how these combinations of LGOs interact with the AOs of central atom (here O-atom).*

Now to make our life easy, let us tabulate the effect of various symmetry operations on those atomic orbital bases. The following table is generated for C_{2v} point group.

Table 10.1: Effect of symmetry operations on various atomic orbitals.

Atomic orbitals	Basis	E	$C_2(z)$	σ_{xz}	σ_{yz}	*Symmetry*
$1s(H_1)$	ψ_1	ψ_1	ψ_2	ψ_1	ψ_2	A_1+B_2
$1s(H_2)$	ψ_2	ψ_2	ψ_1	ψ_2	ψ_1	A_1+B_2
$2s(O)$	ψ_3	ψ_3	ψ_3	ψ_3	ψ_3	A_1
$2p_x(O)$	ψ_4	ψ_4	$-\psi_4$	ψ_4	$-\psi_4$	B_1
$2p_y(O)$	ψ_5	ψ_5	$-\psi_5$	$-\psi_5$	ψ_5	B_2
$2p_z(O)$	ψ_6	ψ_6	ψ_6	ψ_6	ψ_6	A_1

Taking the help of character table and careful inspection of the above transformation table, it is very easy to conclude about the symmetry nature of the valence AOs (ψ_1, ψ_2, ψ_5 and ψ_6) of the central O-atom. Since ψ_3 gives the characters 1 1 1 1 against the four operations, hence it is taken as A_1 symmetry, ψ_4 gives the characters 1 –1 1 –1, hence it is transformed as B_1 symmetry. Similarly, ψ_5 and ψ_6 transform as B_2 and A_1 respectively. The two 1s-orbitals of two H-atoms combine through in-phase and out-of-phase combinations with A_1 and B_2 symmetry respectively (see Sec. 11.1.1)

Therefore, if we summarise, we have a set of two LGOs constructed from two peripheral H-atoms, of A_1 and B_2 symmetry and another set of four AOs of the central O-atom ($2s$, $2p_x$, $2p_y$ and $2p_z$) of A_1, B_1, B_2 and A_1 symmetry. Our next job is to check the possible combinations between these two sets of orbitals.

Here, first observation is that since there is no counterpart for B_1 symmetry of AO of central O-atom (ψ_4) in LGO symmetry, hence this orbital remains as *nonbonding* orbital in water molecule.

Now with same B_2 symmetry, $2p_y$ orbital of O-atom (ψ_5) and one orbital of LGO (out of phase SALC) can combine linearly *in-phase* and *out-of-phase* to give *bonding* and *antibonding* MOs respectively as follows:

$$c_1\psi_5 + c_2\frac{1}{\sqrt{2}}[(\psi_1 - \psi_2)] \quad \text{Or,} \quad 2p_y(O) + [1s(H_1) - 1s(H_2)]$$

and

$$c_1\psi_5 - c_2\frac{1}{\sqrt{2}}[(\psi_1 - \psi_2)] \quad \text{Or,} \quad 2p_y(O) - [1s(H_1) - 1s(H_2)]$$

From the above equations it is evident that bonding has one node and antibonding has three nodes and also we know that energy increases with the number of nodes. We can pictorially represent the bonding and antibonding MOs as shown in Figure 10.5.

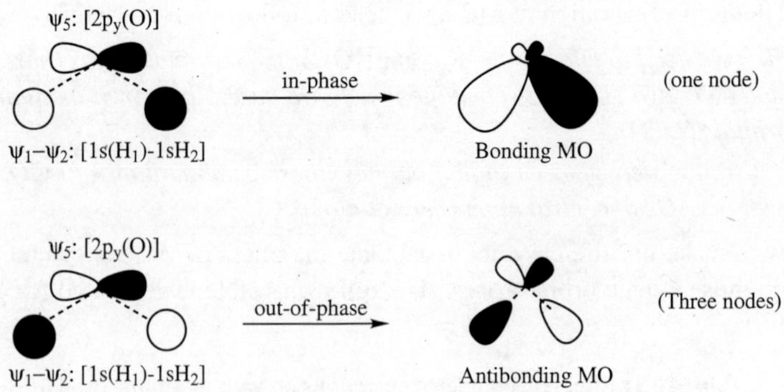

Figure 10.5: Bonding and antibonding MOs.

Now, we have two AOs of A_1 symmetry (ψ_3 and ψ_6) from $2s$(O) and $2p_z$(O) and another of same symmetry from LGO, $\frac{1}{\sqrt{2}}(\psi_1 + \psi_2)$ just constructed through in-phase combination from two $1s$ orbitals of two H-atoms. To construct MO from three AOs of similar symmetry it is common practice primarily to form the combination between AOs of same atom and then consider the combination among these produced AOs with the third AO of another atom. Hence, first consider the *in-phase* and *out-of-phase* combinations of $2s$(O) and $2p_z$(O) orbitals. The mathematical forms of two combinations will be: $2s$(O) + $2p_z$(O) and $2s$(O) - $2p_z$(O), respectively and pictorially it is represented in Figure 10.6.

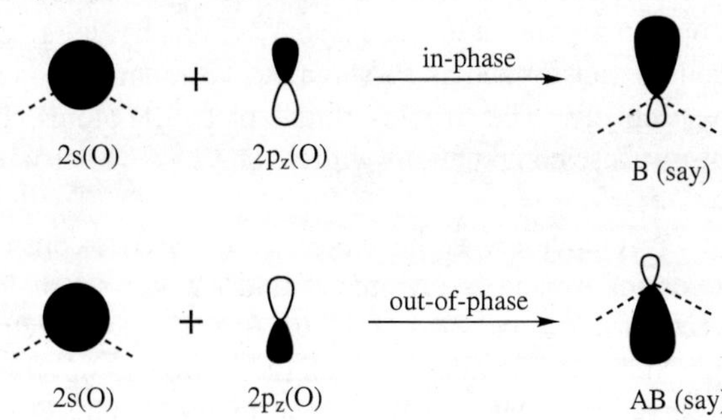

Figure 10.6: *In-phase* and *out-of-phase* combinations of $2s$(O) and $2p_z$(O) orbitals.

From the above diagram, it is evident that the electron density exist far from the two O-H bonds in *in-phase* combination of $2s$(O) and $2p_z$(O) and hence, obviously there will be no efficient interaction or if it happens at all, the interaction would be insignificant with the LGO of same A_1 symmetry (*i.e.*, *in-phase* combination of AOs H_1 and H_2 or $\psi_1 + \psi_2$), hence it will remain as almost nonbonding orbital with slight stabilization of energy. Whereas, in the *out-of-phase* combination of $2s$(O) and $2p_z$(O), the electron density is situated very close (in between) to the two O-H bonds, hence efficient *in-phase* and *out-of-phase* interaction occurs with $\psi_1 + \psi_2$ to result *bonding* and *antibonding* MO as shown in Figure 10.7.

At this stage, one interesting point may be noted. Generally, when two atomic orbitals combine they produce two MOs, one at lower energy than the participating lower energy orbital and the other at higher energy than the participating higher

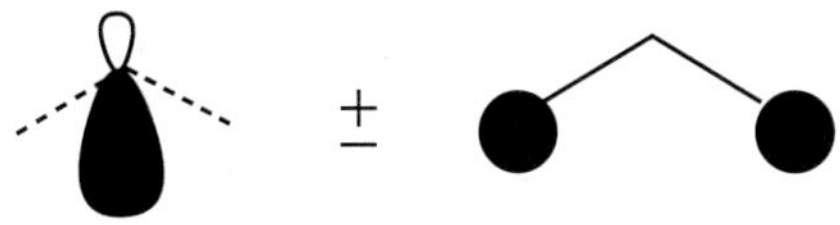

Figure 10.7: Bonding and antibonding MOs.

energy orbital. But, here when $2s$ and $2p_z$ orbital of O-atom get mixed together two form two new orbitals, both are having energies between the energies of pure $2s$ and pure $2p_z$ orbitals. This kind of observation is common when the participating orbitals are from same atom for mixing.

Again, the actual position of the mixed orbitals will depend on the relative contribution of the two participating orbitals. If s-character increases, then there will be lowering of energy and if p-character increases, there will be an increase in energy level.

Hence, keeping all those factors in mind and matching the symmetries during interaction, the overall energy level diagram for water molecule can be constructed as shown in Figure 10.8. The numbering system for the MOs is done in ascending order within the same symmetry.

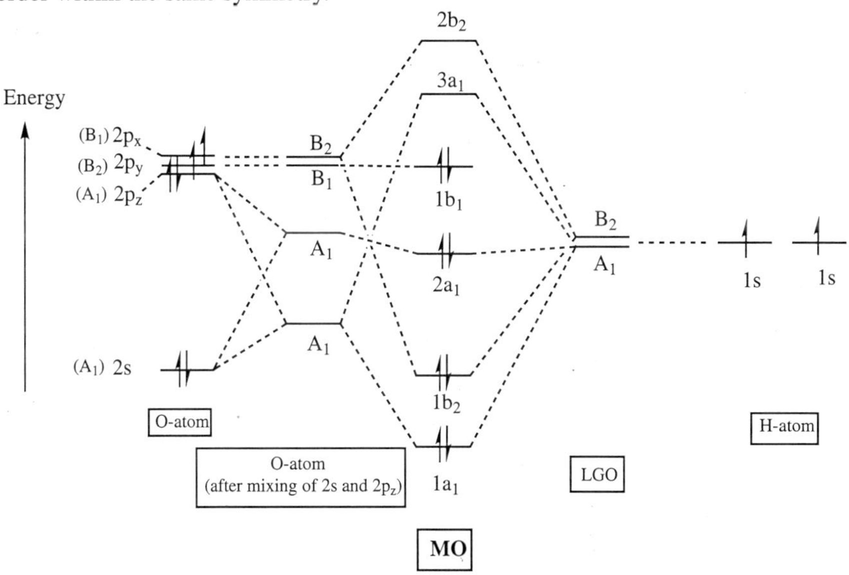

Figure 10.8: MO diagram of water molecule.

10.3.2 CH_4 Molecule

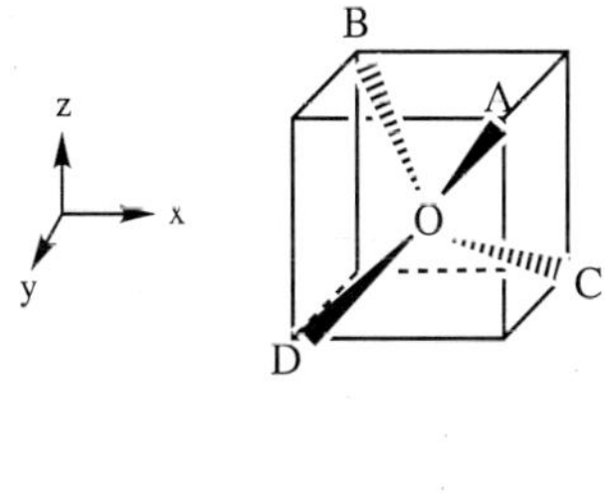

Let us take the second example of methane molecule to treat the problem, through LGO formation, to construct final SALC-MOs. The four H-atoms of this molecule during σ-interaction with central C-atom, are assumed to take tetrahedral configuration (T_d) with four $1s$ orbitals of four H-atoms situated at first, third, sixth and eighth corner of the perfect tetrahedron, are taken as basis, shown in the figure. Now, our job is to form appropriate LGOs through SALC and to find their proper symmetries.

With those bases, we can very easily develop the reducible representation which is shown in the table below. The reducible representation can be reduced to,

$$\Gamma_{SALC} = A_1 + T_2$$

T_d	E	$8C_3$	$3C_2$	$6S_4$	$6\sigma_d$
Γ_{SALC}	4	1	0	0	2

Hence, it is found that there would be total four linear combinations. Out of the four linear combinations, one is totally symmetric, A_1 symmetry and the other three are triply degenerate, T_2 symmetry. Which mathematical combination is of A_1 symmetry and which three combinations are of T_2 symmetry that can be found using projection operator method. But, at this moment we are not familiar with this method.

This method has been discussed in the next chapter taking the same example of methane molecule but to explore molecular vibration. Here, in the same way we can find the exact mathematical form for the four different combinations. The normalized SALC corresponding to A_1 symmetry is found to be a combination of all four $1s$ orbitals combining in same phase, *i.e.*,

$$\psi_1(A_1) = \frac{1}{2}[1s_A + 1s_B + 1s_C + 1s_D]$$

Similarly, corresponding to the T_2 symmetry, triply degenerate combinations are found as:

$$\psi_2(T_2(z)) = \frac{1}{2}[1s_A + 1s_B - 1s_C - 1s_D]$$

$$\psi_3(T_2(y)) = \frac{1}{2}[1s_A - 1s_B - 1s_C + 1s_D]$$

$$\psi_4(T_2(x)) = \frac{1}{2}[1s_A - 1s_B + 1s_C - 1s_D]$$

It is known that C-atom contains both $1s$ and $2s$ orbital and both are of A_1 symmetry. Since $1s$ orbital is a low laying core orbital, hence it may be considered as non-bonding MO (σ_1). The $2s$ orbital (A_1 symmetry) combines with the same symmetry SALC, ψ_1 to form bonding and antibonding MO which have the following forms shwon in Figure 10.9.

$$\sigma_2 = c_1(2s) + c_2\psi_1 \quad \text{and} \quad \sigma_2^* = c_3(2s) - c_4\psi_1$$

The pictorial form of the above two LCAO-MO is shown in Figure 10.9.

σ_2–Bonding σ^*_2–Antibonding

$[2s(C)+\psi_1]$; A_1-**symmetry** $[2s(C)-\psi_1]$; A_1–**symmetry**

Figure 10.9: Bonding and antibonding LCAO-MO.

Here one point may be noted that in the earlier example of water, although we have seen that $2s$ and $2p_z$ got mixed up before interaction with LGOs of H-atom, here, since $2s$ and $2p_z$ of C-atom are not of same symmetry, hence, they do not get mixed up.

In a similar way, three degenerate p-orbitals of C-atom of T_2 symmetry combine with the three SALCs of T_2 symmetry (ψ_2, ψ_3 and ψ_4) produced from four H-atoms. Here each SALC forms bonding MO by matching a $2p$ orbital whose positive lobe is directed between two hydrogen atom wave functions with positive sign and negative lobe of $2p$ orbital is directed between other two hydrogen atoms wave functions with negative sign in a tetrahedron cage. Similarly, for construction of corresponding antibonding MO, the $2p$ lobe will combine with opposite sign. The resulting wave functions along with the pictorial representations for bonding and antibonding MOs (Figure 10.10) are presented below:

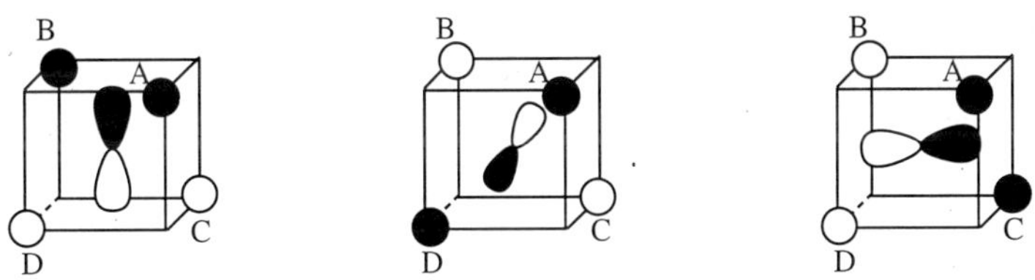

σ_3–Bonding σ_4–Bonding σ_5–Bonding

$[2p_z(C)+\psi_2]$; T_2-**symmetry** $[2p_y(C)+\psi_3]$; T_2-**symmetry** $[2p_x(C)+\psi_4]$; T_2-**symmetry**

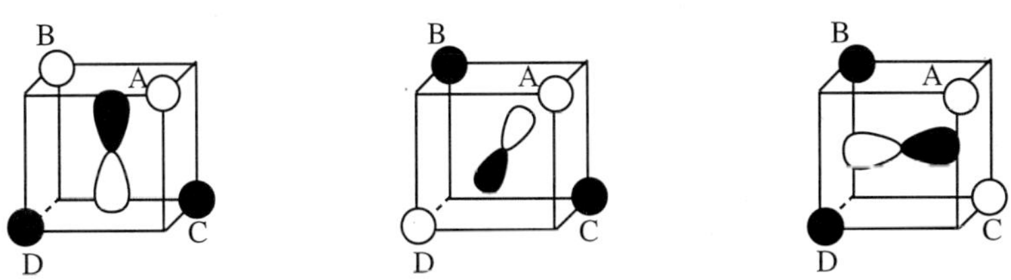

σ_3–Antibonding σ_4–Antibonding σ_5–Antionding

$[2p_z(C)-\psi_2]$; T_2-**symmetry** $[2p_y(C)-\psi_3]$; T_2-**symmetry** $[2p_x(C)-\psi_4]$; T_2-**symmetry**

Figure 10.10: Bonding and antibonding MOs.

$$\sigma_3 = c_5(2p_z) + c_6\psi_2 \quad \text{and} \quad \sigma_3^* = c_7(2p_z) - c_8\psi_2$$

$$\sigma_4 = c_9(2p_y) + c_{10}\psi_3 \quad \text{and} \quad \sigma_4^* = c_{11}(2p_y) - c_{12}\psi_3$$

$$\sigma_5 = c_{13}(2p_x) + c_{14}\psi_4 \quad \text{and} \quad \sigma_5^* = c_{15}(2p_x) - c_{16}\psi_4$$

Here σ_1 is a low laying core orbital and does not take part in bonding process and obviously it remains as nonbonding orbital. Another noticeable point is that σ_3, σ_4 and σ_5 bonding MOs and corresponding σ_3, σ_4 and σ_5 antibonding MOs are degenerate. Following the rules for construction of MO and arrangement of electrons we have the following MO diagram shown in Figure 10.11.

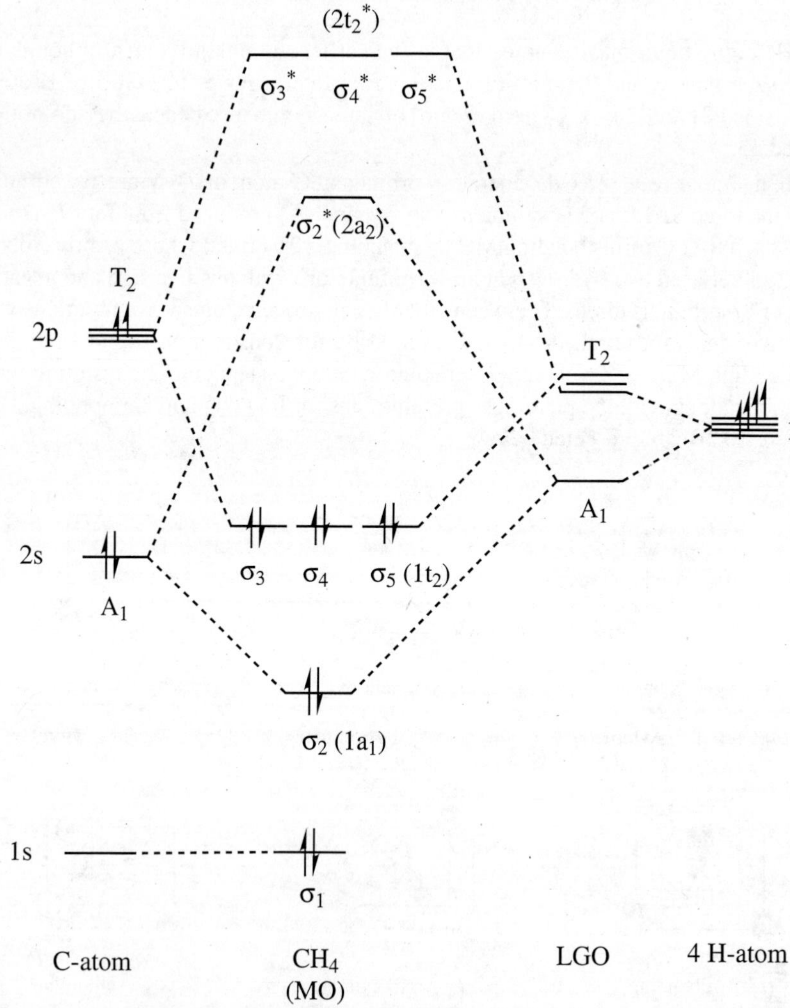

Figure 10.11: MO diagram of tetrahedral methane molecule.

10.3.3 *Staggered*-Ferrocene Molecule, $[(C_5H_5)_2Fe]$

staggered-Ferrocene molecule is an example of organometallic 'sandwich compound', where one metal atom is 'sandwiched' symmetrically between two carbocyclic rings. Although, there are deviation from this general concept, *e.g.*, sometimes the two rings are not symmetrical, or sometimes one carbocyclic ring is present *etc.* Sandwich type of molecules are different from normal ML_n type of metal complex in the sense that, there is a strong electronic interaction between the rings along with the interaction with central metal atom. In ferrocene, the principal, C_5 axis passes through centre of the two cyclopentadienyl rings and the Fe-atom present in-between the rings. Ferrocene molecule may be of two types: one, where two rings are parallel as well as eclipsed position (D_{5h}), and in other case, the two rings are parallel but staggered position with respect to one another (D_{5d}). Although, both the conformers are present in a mixture, but the staggered form being energetically more favourable and hence present as a major component. Here, we will mainly discuss the staggered form of ferrocene. In Figure 10.12, the structure of ferrocene, few important symmetry elements and numbering system of ten p_z-orbitals (according to the numbering system of C-atoms), which are responsible for π- molecular orbital formation, are shown. At this point, one more additional notable feature of the coordinate system for the p_z-orbitals is that the direction of lobes of upper ring and lower ring are exactly opposite, *i.e.*, for example, for C_1, the positive lobe of p_z orbital is downwards, whereas, the direction of lobe in corresponding C_6-atom is upwards. During performance of any symmetry operation on these bases, this point is very important.

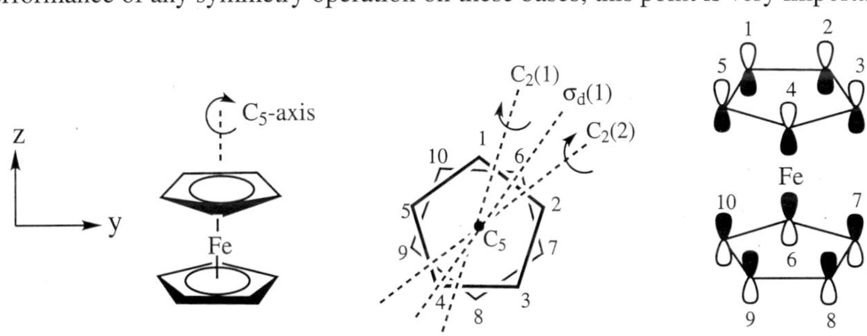

Figure 10.12: Ferrocene molecule, few important symmetry elements, numbering system of p_z-orbitals and coordinate system.

Before going to the discussion of orbital interaction for bond formation, let us have a look on some of the key symmetry elements of *staggered*-ferrocene of D_{5d}-point group. This point group has the following symmetry elements:

$$E, 2C_5, 2C_5^2, 5C_2, i, 2S_{10}^3, 2S_{10}, 5\sigma_d$$

From a top view of the compound, it is seen that the principal axis passes through centre of two rings and Fe-atom as well. There are five C_2 axes perpendicular to the principal axis, which passes through the central Fe-atom. In the picture two such C_2 axes have been shown. One dihedral plane, σ_d has been shown, which is collinear with the C_5-axis and passing through the C-atom of upper ring and mid point of exactly opposite C-C bond of lower ring. Five such dihedral planes can be drawn, as evident from Figure 10.12. The notable interesting point regarding the principal C_5-axis is that C_5^1 and C_5^4 operations belong to a separate class than C_5^2 and C_5^3 operations although, all are originated from C_5 symmetry element. Thus, the first group of operations are clubbed together and designated as $2C_5$ and second group of elements as $2C_5^2$, as evident from the character table of D_{5d} point group. Similar is the case for S_{10} symmetry element, where, $2S_{10}^3$ corresponds to S_{10}^3 and S_{10}^7 operations and $2S_{10}$ corresponds to S_{10}^1 and S_{10}^9 operations. Other operations corresponding to S_{10} element produces other alternative equivalent simple symmetry operations (*e.g.*, $S_{10}^2 \equiv C_5^1, S_{10}^4 \equiv C_5^2, S_{10}^5 \equiv i, S_{10}^6 \equiv C_5^3, S_{10}^8 \equiv C_5^4$ and $S_{10}^{10} \equiv E$).

For construction of π-MO, the sequence of steps are as follows:

Firstly, Find the atomic orbitals of central metal atom (Fe)involved and their associated symmetries easily obtained from character table. Arrange them according to their energies.

Secondly, Construct appropriate SALCs or LGOs from the participating p_z-orbitals of ten C-atoms of the two cyclopentadienyl rings and find their symmetries as well. Arrange them according to their energies as obtained by counting the number of nodal planes of the pictorial representations of the wave functions.

Finally, combine the two sets of orbitals, obtained from AOs of Fe atom and SALC-AO according to their symmetries.

10.3.3.1 Symmetries of Atomic Orbitals Fe-atom

The atomic orbitals for Fe-atom, which take part in bonding process are $3d, 4s$ and $4p$. From the inspection of character table of D_{5d}, it is evident that the symmetries corresponding to those AOs are as follows:

AO		*Symmetry*	AO		*Symmetry*
$4s$	:	a_{1g}	d_{z^2}	:	a_{1g}
$4p_x, 4p_y$	:	e_{1u}	$d_{x^2-y^2}, d_{xy}$	:	e_{2g}
$4p_z$	:	a_{2u}	d_{xz}, d_{yz}	:	e_{1g}

10.3.3.2 Construction of SALCs or LGOs for Cyclopentadienyl rings

Taking ten p_z orbitals of ten C-atoms from the two rings, the reducible representation for D_{5d} point group can easily be tabulated by simply counting the change of basis under each symmetry operation as:

D_{5d}	E	$2C_5$	$2C_5^2$	$5C_2$	i	$2S_{10}^3$	$2S_{10}$	$5\sigma_d$
Γ_{red}	10	0	0	0	0	0	0	2

Since, except under symmetry operations E and $5\sigma_d$, the bases change their position, hence, all other operations produce character zero. Using reduction formula we get,

$$\Gamma_{red} = A_{1g} + A_{2u} + E_{1g} + E_{1u} + E_{2g} + E_{2u}$$

Thus, there would be ten different combinations of ten P_z-orbitals to for AOs (SALCs) corresponding to each symmetry, just obtained above. To project out the exact form of linear combination, the projection operation method is found to be one of the best choice. Although, at this moment, this method is not known to us, has been discussed in details in chapter 11. Readers are suggested to go through the chapter and come back again here. For that a transformation table (Table 10.2) needs to be constructed taking one generating function (say, ϕ_1).

Table 10.2: Transformation table for D_{5d} point group, taking ϕ_1 as generating function.

D_{5d}	E	C_5^1	C_5^4	C_5^2	C_5^3	$C_2(1)$	$C_2(2)$	$C_2(3)$	$C_2(4)$	$C_2(5)$	i	S_{10}^3	S_{10}^7	S_{10}^1	S_{10}^9	
ϕ_1	ϕ_1	ϕ_2	ϕ_5	ϕ_3	ϕ_4	ϕ_6	ϕ_7	ϕ_8	ϕ_9	ϕ_{10}	ϕ_8	ϕ_7	ϕ_9	ϕ_6	ϕ_{10}	cont..

$\sigma_d(1)$	$\sigma_d(2)$	$\sigma_d(3)$	$\sigma_d(4)$	$\sigma_d(5)$
ϕ_1	ϕ_2	ϕ_3	ϕ_4	ϕ_5

Applying projection operation on A_{1g} symmetry to find the linear combination taking ϕ_1 as generating function, the linear combination corresponding to A_{1g} symmetry:

$$\begin{aligned} {}^{A_{1g}}P_{\phi_1} &= \sum[\chi_R(A_{1g}).R\phi_1] \\ &= [(1\times\phi_1)+(1\times\phi_2)+(1\times\phi_5)+(1\times\phi_3)+(1\times\phi_4)+(1\times\phi_6)+(1\times\phi_7) \\ &\quad +(1\times\phi_8)+(1\times\phi_9)+(1\times\phi_{10})+(1\times\phi_7)+(1\times\phi_9)+(1\times\phi_6) \\ &\quad +(1\times\phi_{10})+(1\times\phi_1)+(1\times\phi_2)+(1\times\phi_3)+(1\times\phi_4)+(1\times\phi_5) \\ &= 2(\phi_1+\phi_2+\phi_3+\phi_4+\phi_5+\phi_6+\phi_7+\phi_8+\phi_9+\phi_{10}) \\ &= \frac{1}{\sqrt{10}}(\phi_1+\phi_2+\phi_3+\phi_4+\phi_5+\phi_6+\phi_7+\phi_8+\phi_9+\phi_{10});\ \text{After normalization.} \end{aligned}$$

If the same exercise is carried out taking some other ϕ, same result would be obtained, hence, the combination corresponding to the symmetry, A_{1g} is

$$\Psi_{A_{1g}} = \frac{1}{\sqrt{10}}(\phi_1+\phi_2+\phi_3+\phi_4+\phi_5+\phi_6+\phi_7+\phi_8+\phi_9+\phi_{10})$$

This is basically a combination where all the p_z orbitals of both the rings have *in-phase* combination mode. Hence, obviously there must be another combination where the two rings combine in an out-of-phase fashion. If we try the same procedure of projection operation technique on A_{2u} symmetry, taking ϕ_1 as generating function, we get

$$\Psi_{A_{2u}} = \frac{1}{\sqrt{10}}(\phi_1+\phi_2+\phi_3+\phi_4+\phi_5-\phi_6-\phi_7-\phi_8-\phi_9-\phi_{10})$$

Similarly, other remaining eight linear combinations can be constructed using projection operation technique and can be written down as follows:

$$\begin{aligned} \Psi_{E_{1g}}(1) = \frac{1}{\sqrt{5}}[&(\phi_1+\phi_6)\,cos(\frac{2\pi}{5})+(\phi_2+\phi_7)\,cos2(\frac{2\pi}{5})+(\phi_3+\phi_8)\,cos2(\frac{2\pi}{5}) \\ &+(\phi_4+\phi_9)\,cos(\frac{2\pi}{5})+(\phi_5+\phi_{10})] \end{aligned}$$

$$\Psi_{E_{1g}}(2) = \frac{1}{\sqrt{5}}[(\phi_1+\phi_6)\, sin(\frac{2\pi}{5}) + (\phi_2+\phi_7)\, sin2(\frac{2\pi}{5}) - (\phi_3+\phi_8)\, sin2(\frac{2\pi}{5}) - (\phi_4+\phi_9)\, sin(\frac{2\pi}{5})]$$

$$\Psi_{E_{1u}}(1) = \frac{1}{\sqrt{5}}[(\phi_1-\phi_6)\, cos(\frac{2\pi}{5}) + (\phi_2-\phi_7)\, cos2(\frac{2\pi}{5}) + (\phi_3-\phi_8)\, cos2(\frac{2\pi}{5}) + (\phi_4-\phi_9)\, cos(\frac{2\pi}{5}) + (\phi_5-\phi_{10})]$$

$$\Psi_{E_{1u}}(2) = \frac{1}{\sqrt{5}}[(\phi_1-\phi_6)\, sin(\frac{2\pi}{5}) + (\phi_2-\phi_7)\, sin2(\frac{2\pi}{5}) - (\phi_3-\phi_8)\, sin2(\frac{2\pi}{5}) - (\phi_4-\phi_9)\, sin(\frac{2\pi}{5})]$$

$$\Psi_{E_{2g}}(1) = \frac{1}{\sqrt{5}}[(\phi_1+\phi_6)\, cos2(\frac{2\pi}{5}) + (\phi_2+\phi_7)\, cos(\frac{2\pi}{5}) + (\phi_3+\phi_8)\, cos(\frac{2\pi}{5}) + (\phi_4+\phi_9)\, cos2(\frac{2\pi}{5}) + (\phi_5+\phi_{10})]$$

$$\Psi_{E_{2g}}(2) = \frac{1}{\sqrt{5}}[(\phi_1+\phi_6)\, sin2(\frac{2\pi}{5}) - (\phi_2+\phi_7)\, sin(\frac{2\pi}{5}) + (\phi_3+\phi_8)\, sin(\frac{2\pi}{5}) - (\phi_4+\phi_9)\, sin2(\frac{2\pi}{5})]$$

$$\Psi_{E_{2u}}(1) = \frac{1}{\sqrt{5}}[(\phi_1-\phi_6)\, cos2(\frac{2\pi}{5}) + (\phi_2-\phi_7)\, cos(\frac{2\pi}{5}) + (\phi_3-\phi_8)\, cos(\frac{2\pi}{5}) + (\phi_4-\phi_9)\, cos2(\frac{2\pi}{5}) + (\phi_5-\phi_{10})]$$

$$\Psi_{E_{2u}}(2) = \frac{1}{\sqrt{5}}[(\phi_1-\phi_6)\, sin2(\frac{2\pi}{5}) - (\phi_2-\phi_7)\, sin2(\frac{2\pi}{5}) + (\phi_3-\phi_8)\, sin(\frac{2\pi}{5}) - (\phi_4-\phi_9)\, sin2(\frac{2\pi}{5})]$$

Thus, we have arrived at the ten different combinations of p_z orbitals along with their respective symmetries. These ten SALCs are also called as LGOs for the two cyclopentadienyl rings. Here, one point must be noted that except the first two combinations, where all the p_z orbitals are involved in the formation of SALCs with a one-to-one correspondence between the two rings, the interaction occurs between two different probable partners with some angle between the upper and lower rings in other eight cases. Although, in *ecliped* form of ferrocene, the corresponding partner p_z-orbitals of the two rings in different SALCs are exactly collinear.

Based on the forms of SALCs originated from possible *in-phase* and *out-of-phase* combinations of p_z-orbitals of two rings of ferrocene and their energies corresponding to those wave functions, the LGOs can be drawn and are shown in Figure 10.13.

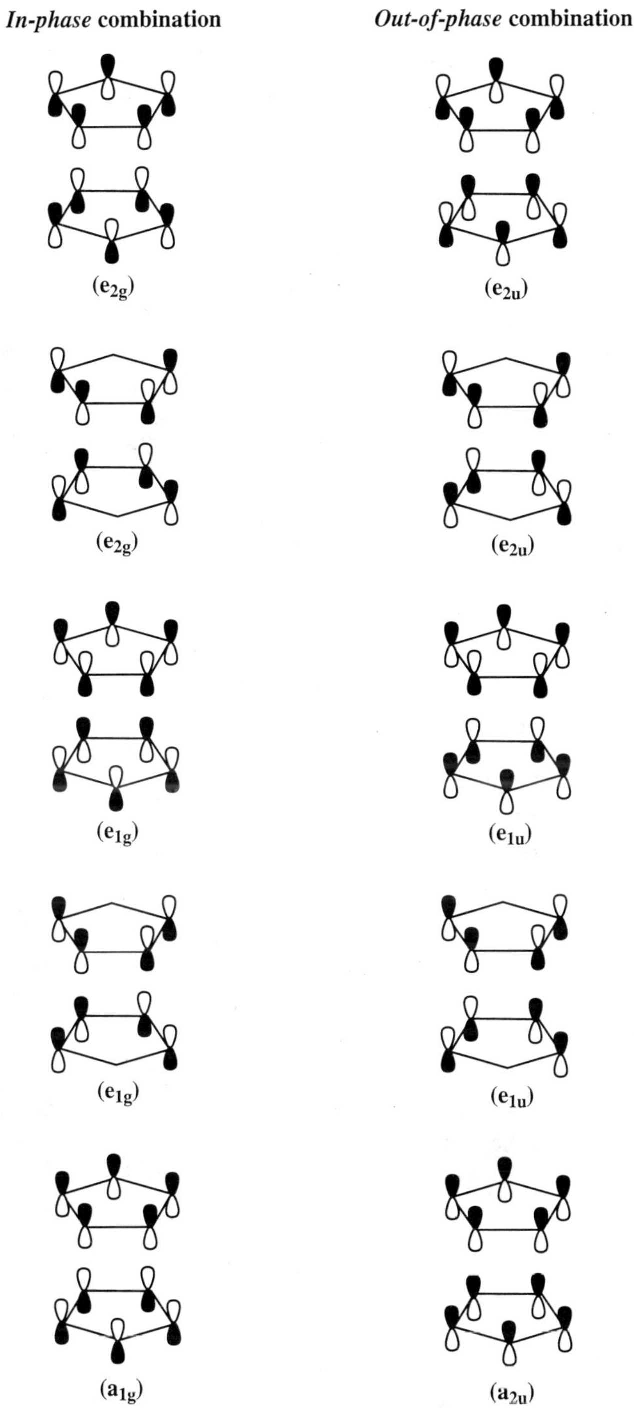

Figure 10.13: Ten SALCs originated from the *in-phase* and *out of phase* combinations of ten p_z-orbitals of two rings of ferrocene along with their symmetries.

10.3.3.3 Construction of Molecular Orbitals

Now our final job is to combine the SALCs originated from the two rings with the AOs of appropriate matching symmetry of Fe-atom. Along with the knowledge of symmetry of those two sets of participating orbitals, knowledge of energy for those orbitals are also very important for proper and efficient formation of MOs. Since, it is well known that the energy ordering of AOs of Fe-atom is $3d < 4s < 4p$, hence, qualitatively they can be arranged as shown in Figure 10.14. Based on the presence of number of nodes present in SALCs gives a qualitative idea about the sequence of energy levels. Since, in *staggered*-ferrocene, the two cyclopentadienyl rings are almost 4Å apart from one another, hence, there is almost negligible interaction between them. As a result, the SALCs formed from *in-phase* or *out-of-phase* combination between the two rings (or *g* and *u* type of orbitals) have hardly any effect on the energy levels. Thus, energy levels corresponding symmetric and antisymmetric combinations have essentially identical energies. Considering those factors, the ordering of energy levels of SALCs are: $a_{1g} \approx a_{2u} < e_{1g} \approx e_{1u} < e_{2g} \approx e_{2u}$.

After the arrangement of orbitals of both the participants (LGOs and Fe-AOs) according to their energies, few points must be kept in mind for symmetry matching as well, for construction MOs.

1. There is no parallel orbital in Fe-atom of e_{2u} symmetry, as present in SALCs. Hence, this orbital is going to remain as non-interacting with metal orbital and emerged as completely ligand-centred MO.
2. $4s$ and d_{z^2} orbitals of Fe combine with similar a_{1g} symmetry SALC of π-LGO to form three MOs of a_{1g} symmetry, namely, $1a_{1g}$, $2a_{1g}$ and $3a_{1g}$.
3. a_{2u}-symmetry LGO and $4p_z$ orbital of Fe-atom combine to one bonding and one antibonding MOs of a_{2u}-symmetry.
4. Doubly degenerate e_{1g} of LGO and metal d_{xz} and d_{yz} of same symmetry combines to generate two degenerate (total four MOs) e_{1g}-symmetry MOs.
5. Doubly degenerate e_{1u} of LGO and metal $4p_x$ and $4p_y$ of same symmetry combines to generate two degenerate (total four MOs) e_{1u}-symmetry MOs.
6. Doubly degenerate e_{2g} of LGO and metal d_{xy} and $d_{x^2-y^2}$ undergoes mixing to produce two pairs of e_{2g}-symmetry MOs.

If we keep these ideas in mind, it becomes a very easy job for us to construct the final MO diagram of *staggered*-ferrocene molecule as shown in Figure 10.14.

There are total 18 electrons in the molecule, since the two cyclopentadienyl rings contribute 10 electrons (from 10 p_z-orbitals) and 8 electrons comes from Fe-atom (2 electron in $4s$-orbital and 6 electrons in $3d$-orbitals. If these 18 electrons are accommodated in the low-lying MOs of the molecule, the electronic configuration becomes:

$$(1a_{1g})^2(1a_{2u})^2(1e_{1g})^4(1e_{1u})^4(1e_{2g})^4(2a_{1g})^4$$

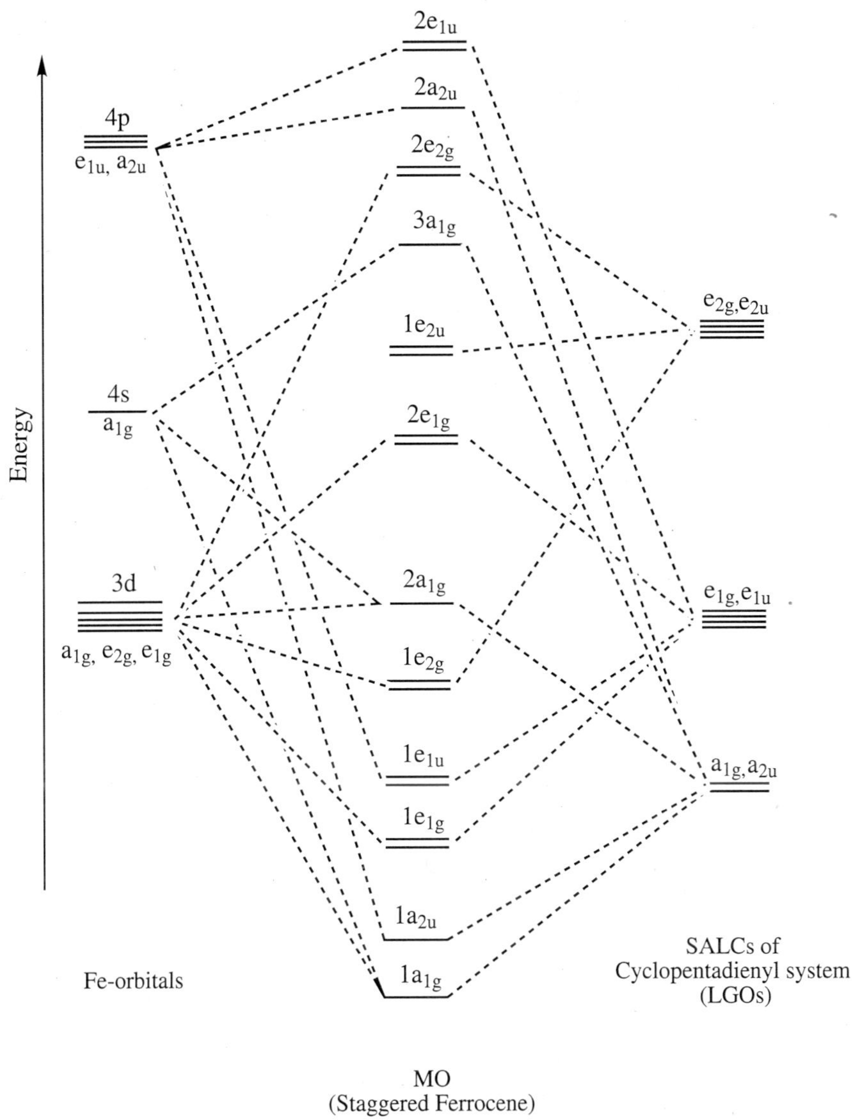

Figure 10.14: MO diagram of *staggered*-ferrocene molecule.

EXERCISES

10.1 From symmetry point of view, find the possible orbitals involved in hybridization for an octahedral, SF_6 molecule.

10.2 In methane molecule, although sp^3 and sd^3, both are symmetrically allowed for hybridization but still the latter case id practically not favourable—explain.

10.3 Construct the MO diagram of *staggered* and *eclipsed*-ferrocene molecule.

10.4 Construct qualitative MO diagram for methane molecule using LGO method.

10.5 Construct qualitative MO diagram for XeF_4 molecule using LGO method.

10.6 Find the SALCs of cyclopentadienyl ring in *staggered*-ferrocene.

10.7 Try to construct LGOs for NH_3 molecule and find their symmetries. Also draw the qualitative MO diagram.

Chapter 11

Projection Operator and its Application in Chemistry

11.1 Formation of Symmetry Adapted Linear Combinations for Molecular Orbitals

It is a technique to generate forms of the SALC (Symmetry adopted linear combination) of atomic orbitals. In addition to its application in MO theory, this technique can be used to find the actual forms of the vibrational modes. So far we have not looked in great depth at topics like Orbitals or Molecular vibrations. So basically some of the earlier applications of group theory will again be revisited applying projection operation technique.

- *We have, for instance, seen that a set of sp^2 hybrids is needed to form a triangular set of bonds but we have not found the wave functions of each orbital.*
- *We have found the symmetry properties of various molecular vibrations and have concluded on their IR and Raman activities. We have not worked out the actual form of each vibrational mode or how actually each and every atom moves in a molecule.*

Symmetry theory can help us to solve all those problems if we extend mathematics a little using the technique, projection operator method. The general form of SALC as:

$$\psi_{SALC} = N[c_1\psi_1 + c_2\psi_2 + \ldots\ldots]$$

Where $\psi_i s$ are the wave functions of atomic orbitals, N is normalization constant.

Let us take one example of XeF_4 (point group, D_{4h}) molecule. We can use four $Xe-F$ bond vectors as the basis for the generation of reducible representation that has been presented in Table 11.1. The reducible representation is further reduced to its components and found to be:

$$\Gamma_{red} = A_{1g} + B_{1g} + E_u$$

Table 11.1: Reducible representation of D_{4h} point group.

D_{4h}	E	$2C_4$	$C_2(=C_4^2)$	$2C_2'$	$2C_2''$	i	$2S_4$	σ_h	$2\sigma_v$	$2\sigma_d$
Γ	4	0	0	2	0	0	0	4	2	0

This indicates that four base vectors can be used either by their sum or difference or their sum/difference to form four linear combinations. One of such combinations transforms as A_{1g} symmetry, another one as B_{1g} and other two transform as E_u symmetry. Similarly using *s*-orbital, three *p*-orbitals or five *d*-orbitals can be used as basis to find similar number of symmetry adapted linear combinations (SALC) corresponding to a specific symmetry (IR of character table). Now the form of the Projection operator is as follow:

$$P_i\psi_j = \sum_R (\chi_{(IR)}.R_{\psi_j})$$

Where P_i is the projection operator for a given IR, i; ψ_j is the chosen basis (or generation function); $\chi_{(IR)}$ is the character in the IR for operator R; and R_{ψ_j} is the effect of performing operation R on the basis ψ_j.

11.1.1 Example 1

Let us take one example of H_2O molecule where two valence shell 1*s* orbitals of two H-atoms are represented by ψ_1 and ψ_2 and one of them are taken as a generating function or basis. Now if we use the projection operation technique, the actual form of linear combination of these two bases can easily be found.

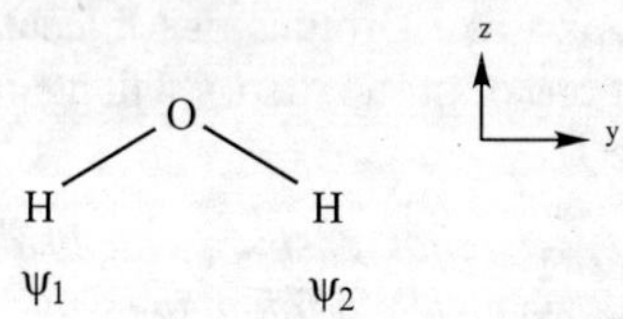

Figure 11.1: Basis or generating function.

Using the two bases (ψ_1 and ψ_2), the reducible representation can be formed as shown in Table 11.2:

Table 11.2: Reducible representation of water molecule taking ψ_1 and ψ_2 as basis.

C_{2v}	E	C_2	$\sigma(xz)$	$\sigma(yz)$
Γ_{red}	2	0	2	0

After reduction, we arrive at the reduced form, $\Gamma_{red} = 2A_1 + B_2$

Now, if we choose orbital ψ_1 as generating function, ψ_1 gets transformed under the symmetry operations of C_{2v} as shown in the following table.

C_{2v}	E	C_2	$\sigma(xz)$	$\sigma(yz)$
ψ_1	ψ_1	ψ_2	ψ_2	ψ_1

Applying projection operator on generating function, ψ_1 for each IR in C_{2v} character table, we get the following combinations.

For IR, A_1, *i.e.*,

$$^{A_1}P_{(\psi_1)} = \sum_R (\chi_{(A_1)}.R_{\psi_1}) \tag{11.1}$$

$$= (1.\psi_1 + 1.\psi_2 + 1.\psi_2 + 1.\psi_1) \tag{11.2}$$

$$= 2\psi_1 + 2\psi_2 \tag{11.3}$$

For IR, A_2; *i.e.*,

$$^{A_2}P_{(\psi_1)} = \sum_R (\chi_{(A_2)}.R_{\psi_1}) \tag{11.4}$$

$$= (1.\psi_1 + 1.\psi_2 + (-1).\psi_2 + (-1).\psi_1) \tag{11.5}$$

$$= 0 \tag{11.6}$$

For IR, B_1; *i.e.*,

$$^{B_1}P_{(\psi_1)} = \sum_R (\chi_{(B_1)}.R_{\psi_1}) \tag{11.7}$$

$$= (1.\psi_1 + (-1).\psi_2 + 1.\psi_2 + (-1).\psi_1) \tag{11.8}$$

$$= 0 \tag{11.9}$$

For IR, B_2; *i.e.*,

$$^{B_2}P_{(\psi_1)} = \sum_R (\chi_{(B_2)}.R_{\psi_1}) \tag{11.10}$$

$$= (1.\psi_1 + (-1).\psi_2 + (-1).\psi_2 + 1.\psi_1) \tag{11.11}$$

$$= 2\psi_1 - 2\psi_2 \tag{11.12}$$

Therefore, it is found that only A_1 and B_2 symmetries could servive after projection operation. These two symmetries can participate in the linear combination process. After normalization, the two SALC's of A_1 and B_2 symmetry are:

$$\frac{1}{\sqrt{2}}(\psi_1 + \psi_2) \text{ and } \frac{1}{\sqrt{2}}(\psi_1 - \psi_2) \text{ respectively.}$$

It is found that the two peripheral atoms (H-atoms) or ligand group orbitals combine one with *in-phase* (*+ve* combination) and another with *out-of-phase* combination (*-ve* combination). Pictorially, they are presented in Figure 11.2.

A_1: $(\psi_1+\psi_2)$ B_2: $(\psi_1-\psi_2)$

Figure 11.2: In-phase and out-of-phase combinations of ligand group orbitals.

Now, using transformation properties of these two combinations under the symmetry operations of C_{2v} point group, it is found that the first combination remains unaltered under all the operations, hence it is of A_1 symmetry type, whereas, the second one produces characters or gets transformed as B_2 symmetry which is clearly evident from the character table.

Final check can be done for the SALCs that have been generated are correct by orthogonality test. The two SALCs must be orthogonal to each other. The functions are orthogonal if the sum of the products of corresponding coefficients is zero *i.e.*, (the product of coefficient of ψ_1 + product of coefficient of ψ_2) = 0. For the above two SALCs of water, this can be verified; $[(1\times 1)+(1\times -1)]=0$

11.1.2 Example 2

In another case, let us take two bond vectors as basis for water molecule for formation of SALC's. Using projection operator, we can find the actual movements of bond vectors during vibration.

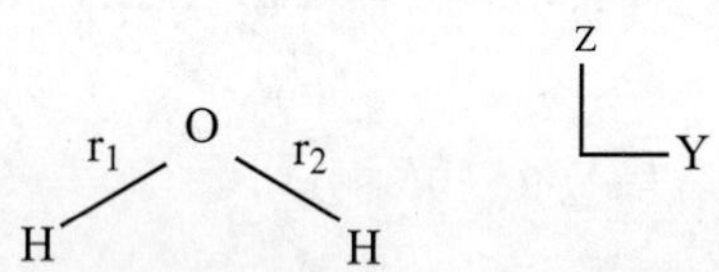

Figure 11.3: Two O-H bond vectors as basis.

Let us consider the two bond vectors, r_1 and r_2 corresponding to the two O–H bonds. The effect of different symmetry operations on the individual basis, or generating functions (r_1 and r_2) are tabulated in Table 11.3:

Table 11.3: Transformation of generating functions.

C_{2v}	E	C_2	$\sigma(xz)$	$\sigma(yz)$
r_1	r_1	r_2	r_2	r_1
r_2	r_2	r_1	r_1	r_2

1. Now, let us take r_1 as generator function and apply projection operator corresponding to $\mathbf{A_1}$ IR and find which combination of r_1 and r_2 forms the basis for A_1.

$$\begin{aligned} {}^{A_1}P_{(r_1)} &= \sum_R [\chi_{(A_1)}.R_{(r_1)}] \\ &= [(1\times r_1)+(1\times r_2)+(1\times r_2)+(1\times r_1)] \\ &= (2r_1+2r_2) \\ &\sim (r_1+r_2) \end{aligned}$$

The coefficients have been ignored since the linear combination of function has to be normalized.

2. Taking r_1 as generator function and applying projection operator corresponding to IR, $\mathbf{A_2}$ is carried out to find which combination of r_1 and r_2 forms the basis for A_2.

$$ {}^{A_2}P_{(r_1)} = \sum_R [\chi_{(A_2)}.R_{(r_1)}] $$

$$= [(1 \times r_1) + (1 \times r_2) + (-1 \times r_2) + (-1 \times r_1)]$$
$$= 0$$

This result indicates that no linear combination of r_1 and r_2 is possible with A_2 symmetry.

3. Taking r_1 as generator function and applying projection operator corresponding to IR,$\mathbf{B_1}$ is carried out to find which combination of r_1 and r_2 forms the basis for B_1.

$$^{B_1}P_{(r_1)} = \sum_R [\chi_{(B_1)} . R_{(r_1)}]$$
$$= [(1 \times r_1) + (-1 \times r_2) + (1 \times r_2) + (-1 \times r_1)]$$
$$= 0$$

Hence, again no linear combination of r_1 and r_2 is possible with B_1 symmetry.

4. Taking r_1 as generator function and applying projection operator corresponding to IR,$\mathbf{B_2}$ symmetry we get;

$$^{B_2}P_{(r_1)} = \sum_R [\chi_{(B_2)} . R_{(r_1)}]$$
$$= [(1 \times r_1) + (-1 \times r_2) + (-1 \times r_2) + (1 \times r_1)]$$
$$= (2r_1 - 2r_2)$$
$$\sim (r_1 - r_2)$$

Therefore, the normalized linear combinations, $\frac{1}{\sqrt{2}}(r_1 + r_2)$ and $\frac{1}{\sqrt{2}}(r_1 - r_2)$ corresponds to O–H stretching of A_1 and B_2 symmetry. The normalization constant for any linear combination is given by $\frac{1}{\sqrt{n}}$; n is the sum of the square of coefficients of individual function. Here ignore the factor "2" in each case as linear combination of function is to be normalized.

From this combination it is very clear that A_1 symmetry corresponds to the movements of two vectors (r_1 and r_2) in phase (*i.e.*, either both are expanding or both are contracting). On the other hand, B_2 symmetry corresponds to the movement of vectors in opposite phase (as r_1 is positive and r_2 is negative; that means when r_1 is expanding, r_2 is contracting or vice versa).

So far we have used r_1 as generator function and found the linear combinations. Now the same exercise can be performed using r_2 as generator function and find the symmetry of different linear combinations.

$$^{A_1}P_{(r_2)} = \sum_R [\chi_{(A_1)} . R_{(r_2)}]$$
$$= [(1 \times r_2) + (1 \times r_1) + (1 \times r_1) + (1 \times r_2)]$$
$$= (2r_1 + 2r_2)$$
$$\sim (r_1 + r_2)$$

$$^{A_2}P_{(r_2)} = \sum_R [\chi_{(A_2)}.R_{(r_2)}] = 0$$

$$^{B_1}P(r_2) = \sum_R [\chi_{(B_1)}.R_{(r_2)}] = 0$$

$$\begin{aligned} ^{B_2}P_{(r_2)} &= \sum_R [\chi_{(B_2)}.R_{(r_2)}] \\ &= [(-1\times r_2)+(1\times r_1)+(1\times r_1)+(-1\times r_2)] \\ &= (2r_1 - 2r_2) \\ &\sim (r_1 - r_2) \end{aligned}$$

Again we see that the two combinations, thus obtained, are of A_1 and B_2 symmetry respectively. The two normalized SALC for water molecule bond vectors are

$$\frac{1}{\sqrt{2}}(r_1+r_2) \text{ and } \frac{1}{\sqrt{2}}(r_1-r_2).$$

Hence, from these form of SALCs, it is very easy to conclude that, during A_1 symmetry type of vibration, both the bond vectors move with same phase *i.e.*, either both bonds are undergoing expansion or both bonds are undergoing compression simultaneously. Hence this mode of vibration must be of symmetric stretching type. Similar way, in B_2 type of vibration, the SALC form indicates that when one bond goes through positive phase (or expansion), other bond goes through compression and vice versa. Hence this mode of vibration must be of type asymmetric stretching.

Final check for the correctness of the generated SALCs can be done by orthogonality test. The two genuine SALCs must be orthogonal to each other. The functions are orthogonal if the sum of the products of corresponding coefficients is zero (*i.e.*, the product of coefficient of r_1 + product of coefficient of $r_2 = 0$). The orthogonality of the SALCs of water can be verified as $(1 \,.\, 1) + (1 \,.\, -1) = 0$.

11.2 Determination of Actual Movements of Atoms during Vibration

11.2.1 Example of BCl_3 Molecule (D_{3h} Point Group)

Let us take the example of BCl_3 molecule (D_{3h}), or ions like CO_3^{-2} or NO_3^- (where degenerate symmetry is involved in SALC). Using projection operator method, here we can find the actual movement of all the atoms.

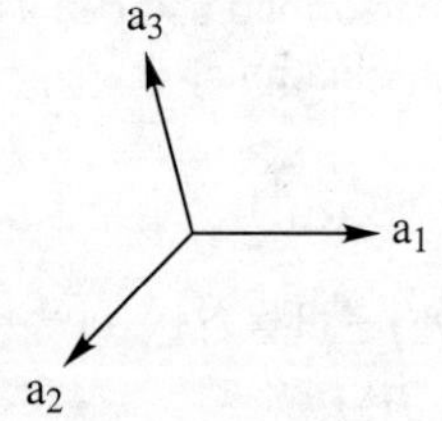

Figure 11.4: Three bond vectors representing stretching mode

Let us take a set of three vectors representing the bond stretching modes as our basis to frame the reducible representation. The reducible representation is shown in the table below:

D_{3h}	E	$2C_3$	$3C_2$	σ_h	$2S_3$	$3\sigma_v$
Γ_{B-Cl}	3	0	1	3	0	1

Using reduction formula and taking the help of D_{6h} character table, we get

$$\Gamma_{B-Cl} = A_1' + E'$$

Hence, for BCl_3, there are three modes of vibrations; one of A_1' symmetry and a degenerate pair of E' symmetry. The projection operation technique will be used to find the movement of the bond vectors for these vibrational modes. For this first we need to find and tabulate (shown below) the nature of transformation against each symmetry operation.

D_{3h}	E	$2C_3$	${C_3}^2$	C_2'	C_2'	C_2'	σ_h	S_3	S_3^2	σ_v	σ_v'	σ_v''
a_1	a_1	a_3	a_2	a_1	a_2	a_3	a_1	a_3	a_2	a_1	a_2	a_3

If we take a_1 as a generating vector for BCl_3 molecule and use the projection operator method as used previously, we get

$$\begin{aligned} {}^{A_1'}P_{(a_1)} &= [(1\times a_1)+(1\times a_3)+(1\times a_2)+(1\times a_1)+(1\times a_2)+(1\times a_3) \\ &\quad +(1\times a_1)+(1\times a_3)+(1\times a_2)+(1\times a_1)+(1\times a_2)+(1\times a_3)] \\ &= 4(a_1+a_2+a_3) \end{aligned}$$

(Terms in each parentheses correspond to various symmetry operations for the D_{3h} point group).

Similarly, linear combination corresponding to E' can be determined as:

$$\begin{aligned} {}^{E'}P_{(a_1)} &= [(2\times a_1)+(-1\times a_3)+(-1\times a_2)+(0\times a_1)+(0\times a_2)+(0\times a_3) \\ &\quad +(2\times a_1)+(-1\times a_3)+(-1\times a_2)+(0\times a_1)+(0\times a_2)+(0\times a_3)] \\ &= (4a_1-2a_2-2a_3) \\ &= 2(2a_1-a_2-a_3) \end{aligned}$$

We do not need to carry out similar exercises with all other IRs of the D_{3h} point group since they do not appear in B-Cl bond stretching mode after reduction and consequently they are not going to produce any linear combination after the projection operation. Therefore, after normalization of the two linear combinations we get:

$$A_1' = \frac{1}{\sqrt{3}}(a_1+a_2+a_3) \text{ and } E' = \frac{1}{\sqrt{6}}(2a_1-a_2-a_3)$$

The magnitudes and signs of the coefficients give the exact form of vibration of the two modes. The A_1' symmetry mode represents the totally symmetric or 'breathing' mode since, all the bond vectors either expand or contract simultaneously to the same extent as shown in Figure 11.5 (panel I). From the composition of E' mode , it is found that if a_1 bond vector stretches two times in magnitute, the contraction of a_2

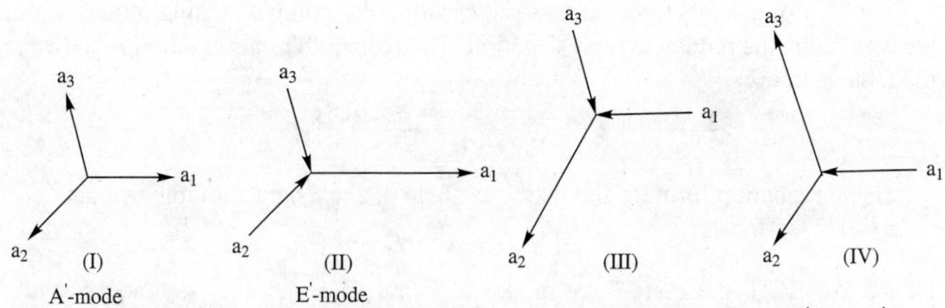

Figure 11.5: Actual direction of bond vectors during stretching vibrations of A_1' and E' symmetry modes.

and a_3 vectors will be just half of the magnitude of a_1 (panel II). The negative signs of a_2 and a_3 vectors indicate the opposite phase in comparison to a_1 vector.

We can further verify the orthogonality of these two SALCs.

$$[1.2+1.(-1)+1.(-1)] = 0.$$

Now, since E' is a degenerate mode, hence we have to find another partner for this E' symmetry mode of vibration which must have identical vibrational frequency with different atomic movements.

To find out another vibrational mode of E' symmetry, the initial temptation must be to select another vector (say a_2) as the generating vector and repeat the above procedure. After completing the same procedure we get the linear combination as $2a_2 - a_3 - a_1$. But this looks very similar to the previous result (panel II), just a rotation through one-third of a turn as shown in the picture (panel III).

But this result is not acceptable, partly because we could go on to generate a third E' mode taking a_3 as generating vector (panel IV) and hence becomes triply degenerate, and also secondly, because the different E' modes would violate the orthogonality criteria (see below). Hence, the latter two combinations (panel III and IV) are not acceptable and would be considered as redundant. Therefore, we have to adopt some other alternative technique.

$$2.(-1)+(-1).2+(-1).(-1) = -3;$$

Hence, $(2a_1 - a_2 - a_3)$ and $(2a_2 - a_3 - a_1)$ are not orthogonal.

Alternative method

If there is a degenerate representation appears in a linear combination, another generating vector is selected at right angles (orthogonal) to the first one (a_1) and is repeated the earlier procedure. Hence, we have the diagram (Figure 11.6) where all the vectors are designated after following vector summation and subtraction rules.

From the nomenclature, the second generating vector is found to be basically $(a_3 - a_2)$ as it lies between a_3 and $-a_2$. Using this $(a_3 - a_2)$ vector as generating vector, we can apply the same projection operator method and complete the table.

Table 11.4: Transformation of second generating vector.

D_{3h}	E	C_3	C_3^2	$C_2(1)$	$C_2(2)$	$C_2(3)$
$(a_3 - a_2)$	$(a_3 - a_2)$	$(a_2 - a_1)$	$(a_1 - a_3)$	$(a_3 - a_2)$	$(a_2 - a_1)$	$(a_1 - a_3)$

contd..	σ_h	S_3	S_3^5	$\sigma_v(1)$	$\sigma_v(2)$	$\sigma_v(3)$
	(a_3-a_2)	(a_2-a_1)	(a_1-a_3)	(a_3-a_2)	(a_2-a_1)	(a_1-a_3)

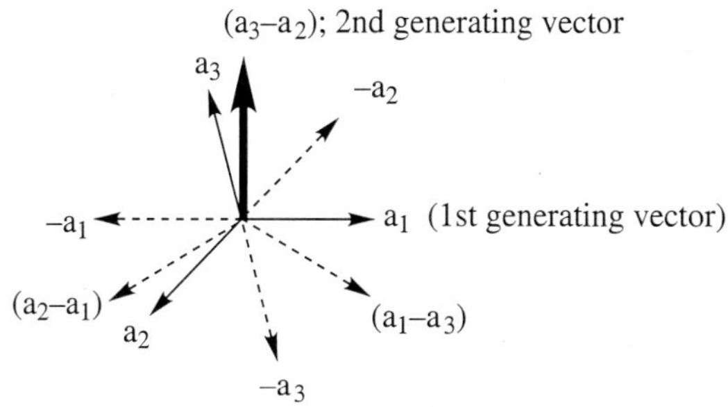

Figure 11.6: Diagram of 2nd generating vector.

We now need to multiply these terms by the characters in the E' representation in the character table to obtain the form of the second E' mode of vibration, *i.e.,*

$$\begin{aligned} {}^{E'}P_{(a_3-a_2)} &= [2\times(a_3-a_2)+(-1)\times(a_2-a_1)+(-1)\times(a_1-a_3)+0+0+0 \\ &\quad +2\times(a_3-a_2)+(-1)\times(a_2-a_1)+(-1)\times(a_1-a_3)+0+0+0] \\ &= 6a_3-6a_2 \end{aligned}$$

After normalization, $E'=\frac{1}{\sqrt{2}}(a_3-a_2)$

In this combination, a_1 is absent and a_2 and a_3 are in opposite phase. Hence the full set of bond stretching vibrations are presented in Figure 11.7.

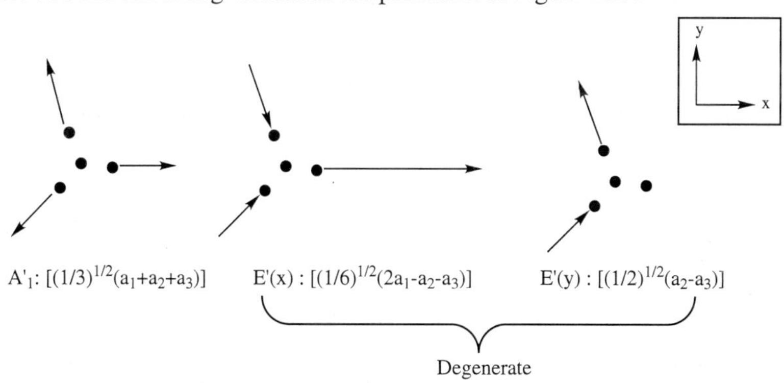

Figure 11.7: Three stretching mode of vibrations corresponding to A_1' and E' symmetries.

The character table of D_{3h} shows that A_1' mode is not IR active but E' modes are active since they belong to the same representation as the x and y directions. Using the above axes shown, we can see that the $[\frac{1}{\sqrt{6}}(2a_1-a_2-a_3)]$ mode of vibration will give rise to an net oscillating dipole moment in the x-direction and hence can be termed as$E'(x)$. Bond 1 (a_1 vector) is oscillating purely along the x-axis; bonds 2 (or a_2 vector) and 3 (a_3 vector) have components along x-axis and y-axis. But the later components for bonds 2 and 3 cancel each other, where as along x-axis component for both the vectors strengthen the a_1 vector. Similarly, $[\frac{1}{\sqrt{2}}(a_2 - a_3)]$ mode gives rise to an oscillating dipole moment along y-direction since the x-components of bond 2 and 3 cancel each other and bond 1 is not taking part at all and hence can be designated as $E'(y)$.

It should be noted that the vibrational modes shown above are, of course, only schematic. When a real molecule or ion vibrates, its center of gravity remains in the same place, so the central atom must move along with the other outer ones. The extent of this movement will depend on the relative masses of the atoms concerned. Thus the NO_3^- ion, containing atoms of similar masses, will behave differently from BCl_3 where the central atom is very light. The symmetries and general form of the vibrational modes are, however, the same in both cases. Now as a final check on the form of our vibrational modes we must ensure that the following two conditions are satisfied.

- *Overall each bond must be used equally.*
- *The functions describing the modes of vibration must be orthogonal to each other.*

These can easily be checked, if we write the coefficients of a_1, a_2 and a_3 for all symmetries in a tabular form, which has been shown in Table 11.5:

Table 11.5: coefficients of a_1, a_2 and a_3 for the three symmetries.

	a_1	a_2	a_3
A_1'	$1/\sqrt{3}$	$1/\sqrt{3}$	$1/\sqrt{3}$
$E'(x)$	$2/\sqrt{6}$	$-1/\sqrt{6}$	$-1/\sqrt{6}$
$E'(y)$	0	$1/\sqrt{2}$	$-1/\sqrt{2}$

If we add squares of the coefficients of a_1, we get: $1/3 + 4/6 = 1$
Similarly, for a_2: $1/3 + 1/6 + 1/2 = 1$ and a_3 : $1/3 + 1/6 + 1/2 = 1$.

Taking the data from the table, the above calculations show that the three bond vectors contribute equally to the total picture of the bond stretching vibrations of the molecule (*first* condition).

Now, to check the *second* condition, we select two modes (say, A_1' and E') and multiply the coefficients of corresponding vectors and after summing up, the result is presented in Table 11.6.

Table 11.6: Test of orthogonality.

	a_1	a_2	a_3	
$A_1' \times E'(x)$	$(\frac{1}{\sqrt{3}} \times \frac{2}{\sqrt{6}})$	$(\frac{1}{\sqrt{3}} \times -\frac{1}{\sqrt{6}})$	$(\frac{1}{\sqrt{3}} \times -\frac{1}{\sqrt{6}})$	SUM =0

The result is found to be *zero*. The result is also found to be *zero* for other two pairs of vibrational modes. Therefore, we can conclude that every mode is orthogonal to each other.

11.2.2 Example of a D_{4h} System

Here instead of projection operator, let us discuss the exact mode of vibrations from character table directly. We will choose the axes such that four vectors point along the coordinate axes. This set of vectors can be used as basis of a representation of D_{4h} which can further be reduced to $A_{1g} + B_{1g} + E_u$.

From character table, A_{1g} (totally symmetric) has the same symmetry as $x^2 + y^2$ which depicts that this mode is a combination of both square terms of x and y-components. All the vectors (a_1, a_3, a_2 and a_4) in x and y-direction must be in their corresponding positive or negative modes simultaneously (let us consider, positive mode as *expansion* of the bond vectors), shown in Figure 11.8. Hence, the normalized vector combination for A_{1g} would be:

$$A_{1g} : 1/2(a_1 + a_2 + a_3 + a_4)$$

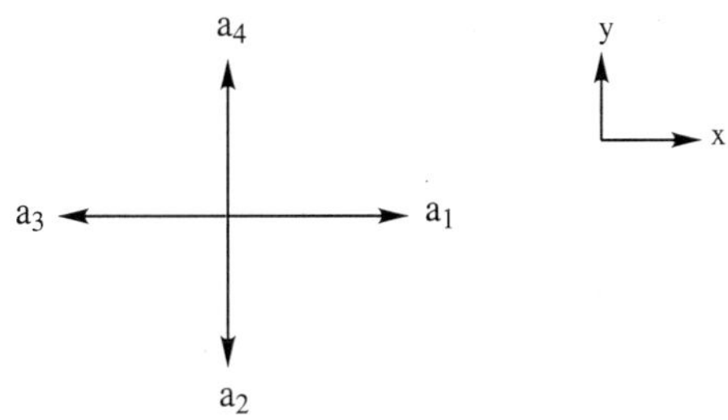

Figure 11.8: Stretching mode of vibration corresponding to A_1' symmetry.

From character table, again it is evident that B_{1g} combination has the same symmetry as $x^2 - y^2$ which means that if x-direction components (a_1 and a_3) are in their corresponding positive mode, y-direction components (a_2 and a_4), vibrate in their corresponding negative mode as shown in Figure 11.9. Hence, the normalized combination for B_{1g} symmetry would be:

$$1/2(a_1 - a_2 + a_3 - a_4)$$

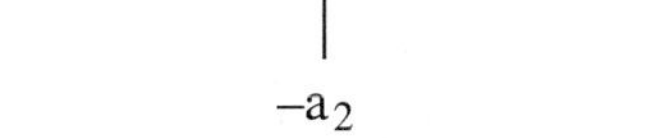

Figure 11.9: Stretching mode of vibration corresponding to B_{1g} symmetry.

Now, in a similar way, combination of the vectors can be constructed for the doubly degenerate E_u symmetry. From the character table, it is seen that it has individual x- and y-direction components. For a planar molecule, stretching vibrational mode can not posses any component in z-direction. One mode would be consisting of only x-component and the other partner mode would be consisting of only y-component. For $E_u(x)$ mode, we have to consider only x-direction bond vectors, *i.e.*, a_1 and a_3

vectors. Since, x is an odd function, hence, out of these two vectors, if a_1 moves in positive phase (say expansion) then a_3 would move in negative phase (then obviously it is contraction). The resulting normalized combination will be:

$$E_u(x) = 1/\sqrt{2}(a_1 - a_3)$$

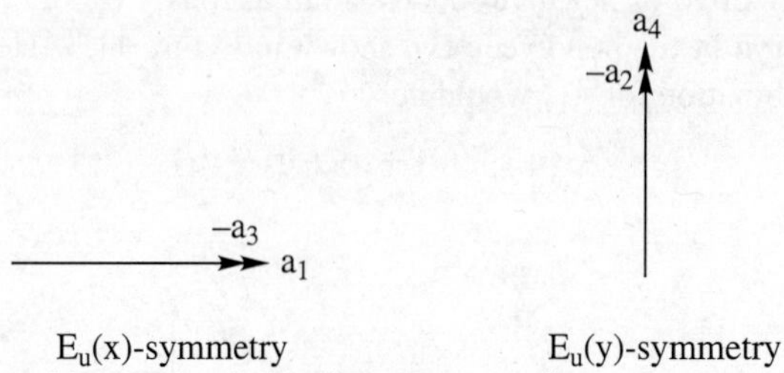

Figure 11.10: Stretching mode of vibrations corresponding to E_u(x) symmetry. (Note that both the vectors in x or y-directions are of equal contribution although, two different arrow heads have been shown only to distinguish the two vectors)

Similarly, E_u(y) would be

$$E_u(y) = 1/\sqrt{2}(a_4 - a_2)$$

The corresponding vibrations have been pictorially presented in Figure 11.10.

From the pictorial diagram of the vector movement for all the four modes, it is evident that A_{1g} and B_{1g} modes of vibration does not cause any change in dipole moments, and hence these two are IR inactive modes. But the degenerate E_u mode of vibration cause obvious oscillating dipole moments in the direction of corresponding axis.

Now, if we consider the coefficients of all the vectors corresponding to all the modes of vibrations, we get the following table:

Table 11.7: Coefficient of vectors corresponding to each symmetry of vibrational mode.

	a_1	a_2	a_3	a_4
A_{1g}	1/2	1/2	1/2	1/2
B_{1g}	1/2	−1/2	1/2	−1/2
E_u(x)	$1/\sqrt{2}$	0	$-1/\sqrt{2}$	0
E_u(y)	0	$-1/\sqrt{2}$	0	$1/\sqrt{2}$

From the coefficients in the table, as we have shown earlier, it can be shown that:

- *Each vector contributes equally to the linear combination (If we square and sum up coefficient a particular vector for all the modes, it results unity.)*
- *Each mode is orthogonal to each other (If we multiply the coefficients of any two modes and then sum up, it becomes zero).*

11.2.3 Example of a T_d System

Let us consider four C–H bond vectors a_1, a_2, a_3 and a_4 along OA, OB, OC and OD respectively. The reducible representation can be formulated as below:

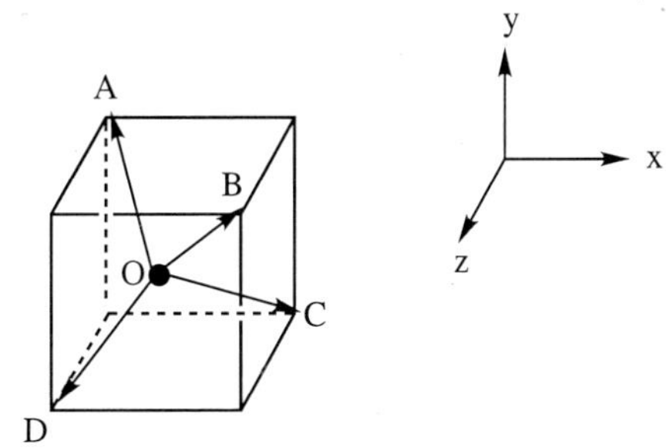

Figure 11.11: T_d geometry with reference axes.

T_d	E	$8C_3$	$3C_2$	$6S_4$	$6\sigma_d$
Γ_{C-H}	4	1	0	0	2

After reducing we get,

$$\Gamma_{C-H} = A_1 + T_2$$

Taking the help of character table of T_d point group and using the projection operator technique we can find the combination of the bond vectors corresponding to A_1 symmetry. Let us start with **a_1 bond vector as a generating vector**. The effect of different symmetry operations of T_d on a_1 can be tabulated as follows:

Table 11.8: Effect of different symmetry operations of T_d on a_1 generating vector.

T_d	E	C_3(A)	C_3(B)	C_3(C)	C_3(D)	C_3^2(A)	C_3^2(B)	C_3^2(C)	
$\mathbf{a_1}$	a_1	a_1	a_4	a_2	a_3	a_1	a_3	a_4	contd...

C_3^2(D)	C_2(x)	C_2(y)	C_2(z)	S_4(x)	S_4(y)	S_4(z)	S_4^3(x)	S_4^3(y)	
a_2	a_2	a_3	a_4	a_4	a_2	a_3	a_2	a_3	contd...

S_4^3(z)	$\sigma_d(AB)$	$\sigma_d(AC)$	$\sigma_d(AD)$	$\sigma_d(BC)$	$\sigma_d(BD)$	$\sigma_d(CD)$
a_4	a_1	a_1	a_1	a_4	a_3	a_2

Taking a_1 as generating vector and applying projection operator formula for $\mathbf{A_1}$ symmetry we get,

$$\begin{aligned} {}^{A_1}P_{(a_1)} &= [(1\times a_1)+(1\times a_1)+(1\times a_4)+(1\times a_2)+(1\times a_3)+(1\times a_1)+(1\times a_3) \\ &\quad +(1\times a_4)+(1\times a_2)+(1\times a_2)+(1\times a_3)+(1\times a_4)+(1\times a_4)+(1\times a_2) \\ &\quad +(1\times a_3)+(1\times a_2)+(1\times a_3)+(1\times a_4)+(1\times a_1)+(1\times a_1)+(1\times a_1) \\ &\quad +(1\times a_4)+(1\times a_3)+(1\times a_2)] \\ &= 6a_1+6a_2+6a_3+6a_4 \\ &= 6(a_1+a_2+a_3+a_4) \end{aligned}$$

After normalization,

$$A_1 = \frac{1}{2}(a_1+a_2+a_3+a_4)$$

Similarly, with a_2, a_3 and a_4 bond vectors considering as generating vector, this ended up with same result. Hence, in this mode of vibration all the four bonds either expand or contract simultaneously and equally, *i.e.*, the vibrational mode associated with this symmetry is IR inactive breathing mode and pictorially this A_1 mode can be expressed as shown in Figure 11.12. Now, applying projection operator formula

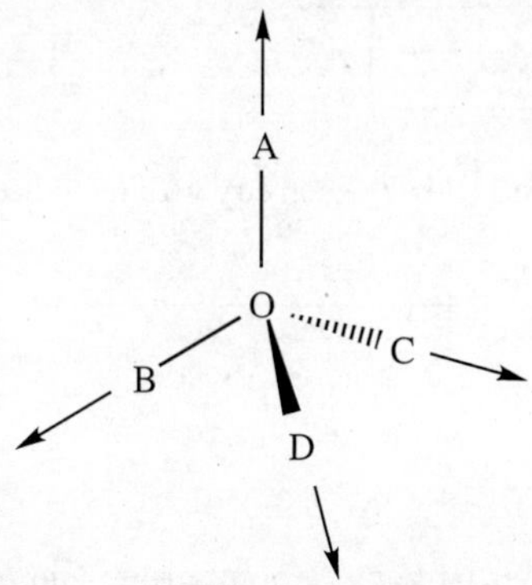

Figure 11.12: Pictorial representation of A_1 mode of vibration for T_d geometry

for $\mathbf{T_2}$ symmetry considering a_1 as generating vector we get,

$$^{T_2}P_{(a_1)} = 6a_1 - 2a_2 - 2a_3 - 2a_4 \tag{11.13}$$

$$= 2(3a_1 - a_2 - a_3 - a_4) \tag{11.14}$$

After normalization,

$$1/\sqrt{12}(3a_1 - a_2 - a_3 - a_4)$$

Again, in a similar way, if the same technique is carried out taking a_2, a_3 and a_4 as generating function, the result would be:

$$^{T_2}P_{(a_2)} = 1/\sqrt{12}(-a_1 + 3a_2 - a_3 - a_4) \tag{11.15}$$

$$^{T_2}P_{(a_3)} = 1/\sqrt{12}(-a_1 - a_2 + 3a_3 - a_4) \text{ and} \tag{11.16}$$

$$^{T_2}P_{(a_4)} = 1/\sqrt{12}(-a_1 - a_2 - a_3 + 3a_4) \tag{11.17}$$

Here few interesting points which must be noted. *Firstly,* $\mathbf{T_2}$, in spite of being a triply degenerate mode, *four* SALCs are produced for this vibration. *Secondly,* after careful inspection, these four SALCs are found to be linearly dependent since,

$$[^{T_2}P_{(a_1)} + ^{T_2}P_{(a_2)} + ^{T_2}P_{(a_3)} + ^{T_2}P_{(a_4)}] = 0$$

It is known that a set of SALCs must always be linearly independent from each other, meaning none of them can be expressed as a linear combination of the others.

Hence, this set of four SALCs are not correctly determined. We have to find out the proper combination of SALCs to produce three functions, $T_2(1)$, $T_2(2)$ and $T_2(3)$

which will follow the necessary criteria. With a little trial-and –error manipulation, we can find that the following three normalized functions can fulfil the criteria of proper SALC.

$$T_2(1) = \frac{1}{2}(a_1 + a_2 - a_3 - a_4) \quad [\text{combination of } {}^{T_2}P_{(a_1)} + {}^{T_2}P_{(a_2)}] \quad (11.18)$$

$$T_2(2) = \frac{1}{2}(a_1 - a_2 - a_3 + a_4) \quad [\text{combination of } {}^{T_2}P_{(a_2)} + {}^{T_2}P_{(a_3)}] \quad (11.19)$$

$$T_2(3) = \frac{1}{2}(a_1 - a_2 + a_3 - a_4) \quad [\text{combination of } {}^{T_2}P_{(a_2)} + {}^{T_2}P_{(a_4)}] \quad (11.20)$$

Since, the combination of ${}^{T_2}P_{(a_3)} + {}^{T_2}P_{(a_4)}$ gives the same result as obtained by combining ${}^{T_2}P_{(a_1)} + {}^{T_2}P_{(a_2)}$, hence it is not considered as a separate combination. These three SALCs, ($T_2(1)$, $T_2(2)$ and $T_2(3)$) are orthogonal to each other. From the SALCs above, if the bond vectors are drawn considering the direction (+ve or --ve sign) and magnitude of coefficient of each bond vectors, the following pictures can be generated generated, presented in Figure 11.13.

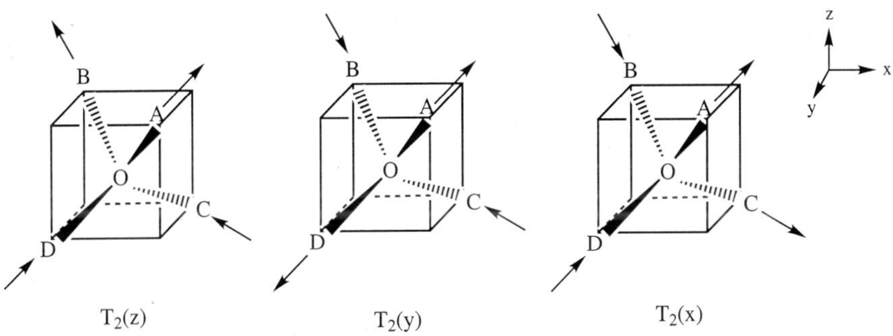

Figure 11.13: Pictorial representation of T_2 mode of vibrations

As for example, in $T_2(1)$ combination, a_1 and a_2 vectors have *(+)ve* sign, where as a_3 and a_4 have *(–)ve* sign. This indicates that when OA and OB bond vectors are undergoing expansion, OC and OD bond vectors are undergoing compression simultaneously. From reference Cartesian coordinates, it is pretty clear that the net change of the direction of participating vectors would be along z-direction. Hence, $T_2(1)$ mode can also be designated as $T_2(z)$. Similarly, $T_2(2)$ and $T_2(3)$ modes can be designated as $T_2(y)$ and $T_2(x)$ respectively. We can verify the two basic criteria for function of SALCs.

- *Overall each bond vectors is used equally in different possible vibrations.*
- *The functions describing the modes of vibration must be orthogonal to each other.*

These two criteria can be checked easily, if we write the matrix of coefficients of a_1, a_2, a_3 and a_4 in Table 11.9.

Table 11.9: Coefficient of vectors corresponding to each symmetry of vibrational mode.

T_d	a_1	a_2	a_3	a_4
A_1	1/2	1/2	1/2	1/2
T_2(x)	1/2	1/2	–1/2	–1/2
T_2(y)	1/2	–1/2	–1/2	1/2
T_2(z)	1/2	–1/2	1/2	–1/2

If we add squares of the coefficients of a_1 we get: $\frac{1}{4}+\frac{1}{4}+\frac{1}{4}+\frac{1}{4}=1$

Same exercises can be done for other vectors, a_2, a_3 and a_4 and found to be *unity*. This shows that all the four bonds contribute equally to the total picture of the bond stretching vibrations of the molecule.

To check the 2nd condition, we select two modes (say, A_1 and T_2(x)) and multiply the coefficients of each vector and sum the result as shown the table below:

Table 11.10: Table for orthogonality test.

	a_1	a_2	a_3	a_4	
$A_1 \times T_2(x)$	$(1/2\times 1/2)$	$(1/2\times 1/2)$	$(1/2\times(-1/2))$	$(1/2\times(-1/2))$	SUM = 0
$T_2(x)\times T_2(y)$	$(1/2\times 1/2)$	$(1/2\times(-1/2))$	$(1/2\times(-1/2))$	$(1/2\times 1/2)$	SUM = 0

The result would be zero for any pair of vibrational modes in the list. Therefore, it can be concluded that every mode is orthogonal to each other. Hence, the final conclusion is that the four SALCs are correctly determined.

11.2.4 Example of a O_h System

Let us consider one AB_6 molecule which is octahedral in geometry. Different ways of representation of the geometry have been provided in Figure 11.14. Consider six bond vectors a_1, a_2, a_3, a_4, a_5 and a_6 along each A-B bond as bases and try to find the reducible representation that has been presented in Table 11.11. Here, only the stretching of different bonds are considered, other kind of bond deformation movements (like bending *etc.*), are not considered.

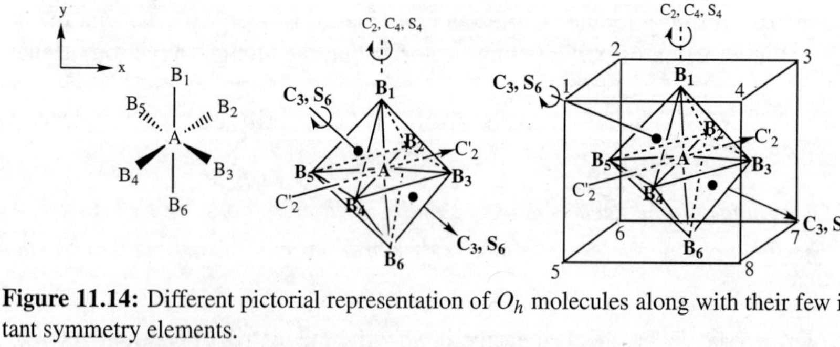

Figure 11.14: Different pictorial representation of O_h molecules along with their few important symmetry elements.

After reduction,

$$\Gamma_{(A-B)bond} = A_{1g} + E_g + T_{1u}$$

Table 11.11: Reducible representation for O_h molecule.

O_h	E	$8C_3$	$6C_2'$	$6C_4$	$3C_2$	i	$6S_4$	$8S_6$	$3\sigma_h$	$6\sigma_d$
$\Gamma_{(A-B)\ bond}$	6	0	0	2	2	0	0	0	4	2

Now, to carry out projection operation, we need to construct the transformation table taking into account total 48 symmetry operations of O_h point group, taking a_1 as generating function, for each of the symmetries found above. The numbers in the parentheses in Table 11.12 indicate the numberining system of the cube. Let us start with A_{1g} symmetry.

Table 11.12: Transformation table of O_h geometry on a_1 generating vector.

O_h	E	$C_3(1,7)$	$C_3(2,8)$	$C_3(3,5)$	$C_3(4,6)$	$C_3^2(1,7)$	$C_3^2(2,8)$	$C_3^2(3,5)$	$C_3^2(4,6)$	
$\mathbf{a_1}$	a_1	a_4	a_5	a_2	a_3	a_5	a_2	a_3	a_4	contd...

$C_2'(1)$	$C_2'(2)$	$C_2'(3)$	$C_2'(4)$	$C_2'(5)$	$C_2'(6)$	$C_4(1)$	$C_4(2)$	$C_4(3)$	$C_4^3(1)$	$C_4^3(2)$	$C_4^3(3)$	
a_6	a_6	a_2	a_4	a_3	a_5	a_1	a_5	a_2	a_1	a_3	a_4	contd...

$C_2(1)$	$C_2(2)$	$C_2(3)$	i	$S_4(1)$	$S_4(2)$	$S_4(3)$	$S_4^3(1)$	$S_4^3(2)$	$S_4^3(3)$	$S_6(1)$	$S_6(2)$	$S_6(3)$	$S_6(4)$	
a_1	a_6	a_6	a_6	a_6	a_3	a_4	a_6	a_5	a_2	a_3	a_4	a_5	a_2	contd..

$S_6^5(1)$	$S_6^5(2)$	$S_6^5(3)$	$S_6^5(4)$	$\sigma_h(1)$	$\sigma_h(2)$	$\sigma_h(3)$	$\sigma_d(1)$	$\sigma_d(2)$	$\sigma_d(3)$	$\sigma_d(4)$	$\sigma_d(5)$	$\sigma_d(6)$
a_2	a_3	a_4	a_5	a_6	a_1	a_1	a_1	a_1	a_3	a_5	a_2	a_4

Here, we are not showing the same procedure as done earlier, but can be completed the projection operation process by multiplying the corresponding characters for each symmetry operations of A_{1g} symmetry (see O_h character table) with these corresponding transformed a_1 bond vectors under each symmetry operation (shown in Table 11.11) and then after summing up, we get:

$$^{A_{1g}}P_{(a_1)} = 8a_1 + 8a_2 + 8a_3 + 8a_4 + 8a_5 + 8a_6$$

Or, After normalization, it can be rewritten as,

$$\Omega_1(A_{1g}) = 1/\sqrt{6}(a_1 + a_2 + a_3 + a_4 + a_5 + a_6)$$

If we carry out projection operation on A_{1g} symmetry taking a_2 or a_3 or any of the other remaining three bond vector as a generating function, same result is obtained. Hence, we do not need to carry out those similar exercises.

Next, carry out the projection operation for E_g symmetry taking a_1 as generating function:

$$^{E_g}P_{(a_1)} = 8a_1 - 4a_2 - 4a_3 - 4a_4 - 4a_5 + 8a_6$$

Or, After normalization, it can be rewritten as,

$$\Omega_2(E_g) = 1/2\sqrt{3}(2a_1 - a_2 - a_3 - a_4 - a_5 + 2a_6)$$

This is indeed one SALC, which can be verified from orthogonality test with the previous SALC of A_{1g}, *i.e.*, $\Omega_1(A_{1g})$. Hence, this is the first one of the degenerate pair of SALCs of E_g symmetry. We have to find out the other one. The first spontaneous attempt would be to choose some other generating function (say a_2 or a_3 *etc.*), and complete the projection operation process, which will end up with the form as (taking a_2 as generating function):

$$^{E_g}P_{(a_2)} = -4\mathrm{a}_1 + 8\mathrm{a}_2 - 4\mathrm{a}_3 + 8\mathrm{a}_4 - 4\mathrm{a}_5 - 4\mathrm{a}_6$$

Or, after normalization, it takes the form:

$$\Omega_2'(E_g) = 1/2\sqrt{3}(-\mathrm{a}_1 + 2\mathrm{a}_2 - \mathrm{a}_3 + 2\mathrm{a}_4 - \mathrm{a}_5 - \mathrm{a}_6)$$

Interestingly, although this form of SALC is orthogonal to the SALC, A_{1g}, but not orthogonal to its partner SALC, *i.e.,* $\Omega_2(E_g)$. Therefore, this form of the SALC is not correct, and hence is not acceptable.

To find the partner SALC for $\Omega_2(E_g)$, as we have done earlier, one way is to make a suitable linear combination between $\Omega_2(E_g)$ and $\Omega_2'(E_g)$ with a form like

$$a\Omega_2(E_g) + b\Omega_2'(E_g);$$

where a and b are small integers. We need to find the suitable values of a and b with some trial and error method so that the form of the SALC, thus formed, obeys the two basic criteria of SALCs . The ultimate form is found to be, $\Omega_2(E_g) + 2\Omega_2'(E_g)$, *i.e.,*

$$\Omega_3(E_g) = (2\mathrm{a}_1 - \mathrm{a}_2 - \mathrm{a}_3 - \mathrm{a}_4 - \mathrm{a}_5 + 2\mathrm{a}_6) + (-2\mathrm{a}_1 + 4\mathrm{a}_2 - 2\mathrm{a}_3 + 4\mathrm{a}_4 - 2\mathrm{a}_5 - 2\mathrm{a}_6)$$

Or, after normalization,

$$\Omega_3(E_g) = \frac{1}{2}(\mathrm{a}_2 - \mathrm{a}_3 + \mathrm{a}_4 - \mathrm{a}_5)$$

In a similar way, using projection operation we can try to find the SALCs for the triply degenerate, T_{1u} symmetry and like the previous case, this is not so difficult to find the three SALCs. Here we start projection operation taking a_1 as generating function and we find the SALC as $^{T_{1u}}P_{(a_1)} = 8a_1 - 8a_6$

Or, after normalization,

$$\Omega_4(T_{1u}) = \frac{1}{\sqrt{2}}(\mathrm{a}_1 - \mathrm{a}_6)$$

It is found that this SALC is orthogonal to those found earlier SALCs. Now a careful observation, it reveals that this SALC is a combination of two functions, a_1 and a_6 which are in *trans* position to each other in an octahedral geometry or they can be taken as equivalent functions. Now, if we again perform projection operation on a_2, we get a form of SALC as

$$\Omega_5(T_{1u}) = \frac{1}{\sqrt{2}}(a_2 - a_4).$$

It also fulfills the criteria of a SALC and again it is a combination of two functions or basis vecctors which are at equivalent positions of an octahedron. Then we are left with another trivial job to find the third partner of the triply degenerate SALCs and will have the form as

$$\Omega_6(T_{1u}) = \frac{1}{\sqrt{2}}(a_3 - a_5)$$

Therefore, using projection operation method, the full set of various symmetries corresponding to AB bond stretching movements of an octahedron are:

$$\begin{aligned}
\Omega_1(A_{1g}) &= 1/\sqrt{6}(\mathrm{a}_1 + \mathrm{a}_2 + \mathrm{a}_3 + \mathrm{a}_4 + \mathrm{a}_5 + \mathrm{a}_6) \\
\Omega_2(E_g) &= 1/2\sqrt{3}(2\mathrm{a}_1 - \mathrm{a}_2 - \mathrm{a}_3 - \mathrm{a}_4 - \mathrm{a}_5 + 2\mathrm{a}_6) \\
\Omega_3(E_g) &= \frac{1}{2}(\mathrm{a}_2 - \mathrm{a}_3 + \mathrm{a}_4 - \mathrm{a}_5) \\
\Omega_4(T_{1u}) &= \frac{1}{\sqrt{2}}(\mathrm{a}_1 - \mathrm{a}_6) \\
\Omega_5(T_{1u}) &= \frac{1}{\sqrt{2}}(\mathrm{a}_2 - \mathrm{a}_4) \\
\Omega_6(T_{1u}) &= \frac{1}{\sqrt{2}}(\mathrm{a}_3 - \mathrm{a}_5)
\end{aligned}$$

11.3 Determination of π-MO and Corresponding Transition Activity

Huckel, in 1931, developed one approximate method for solving the Schrodinger equation to obtain the wave function and energy of π-type MOs of conjugated molecules. This method basically uses the linear variation method and also makes the following assumptions:

- In a conjugated system, the π-electrons are moving in the field of nuclei, core electrons and σ-bond framework. The σ-electrons and π-electrons can be considered as independent to one another and hence can be considered separately.
- Each of the conjugated atoms (labelled 1, 2, 3…) in the molecule contributes one *p*-type atomic orbital perpendicular to the plane of the σ-bond framework. If the atomic orbital wave functions are $\phi_1, \phi_2, \phi_3, \ldots.\phi_n$ (which are obviously not eigen functions of the molecular Hamiltonian, **H**) then

$$\begin{aligned}
\langle\phi_i|\mathbf{H}|\phi_j\rangle = \mathbf{H}_{ij} &= \beta_{ij} \ \text{if} i \text{and} j \text{are adjacent atoms} \\
&= \alpha_i \ \text{if} i = j \\
&= 0 \ \text{if} i \neq j \text{and} i \text{and} j \text{are non-adjacent atoms}
\end{aligned}$$

$$\begin{aligned}
\text{Again,} \langle\phi_i|\phi_j\rangle = \mathbf{S}_{ij} &= 1 \ \text{if} i = j \text{are adjacent atoms} \\
&= 0 \ \text{if} i \neq j
\end{aligned}$$

Here, α_i is called coulomb integral, means the energy of a π-electron if it is localized in atom *i*, but feels the field of the nuclei and electrons (Since the Hamiltonian operator is for the whole molecule and not of the atom i) and β_{ij} is resonance integral which means the energy of a π-electron due to its exchange between the atoms i and j. Now with this concept, we can take one simple π-molecular system, ethylene.

11.3.1 Ethylene (D_{2h})

The methodology is that, starting with a linear combination of the p-type wave functions (ϕ_i) as a trial function and applying the variation method, one end up with a Secular determinant equation. The roots of the equation are the energies of the π-MOs.

Let a trial wave function of a π-MO is:

$$\Psi = c_1\phi_1 + c_2\phi_2$$

Where c_1 and c_2 are combination coefficients to be determined by variation method, and ϕ_1 and ϕ_2 functions represent p_z type atomic orbitals of ethylene carbon atoms labeled as 1 and 2. XY is the molecular plane.

The Schrodinger equation for the π-electronic part is $\mathbf{H}\Psi = E\Psi$

If we multiply both side by ψ^* and integrate over whole space we get,

$$\langle\phi_i|\mathbf{H}|\phi_j\rangle = E\langle\phi_i|\phi_j\rangle$$

Or

$$E = \frac{\langle\psi_i|\mathbf{H}|\psi_j\rangle}{\langle\psi_i|\psi_j\rangle}$$

$$= \frac{c_1\phi_1 + c_2\phi_2|H|c_1\phi_1 + c_2\phi_2}{c_1\phi_1 + c_2\phi_2|c_1\phi_1 + c_2\phi_2}$$

$$= \frac{c_1^2H_{11} + 2c_1c_2H_{12} + c_2^2H_{12}}{c_1^2S_{11} + 2c_1c_2S_{12} + c_2^2S_{12}}$$

$$= \frac{c_1^2\alpha_1 + 2c_1c_2\beta_{12} + c_2^2\alpha_2}{c_1^2 + c_2^2}$$

Or

$$c_1^2(\alpha_1 - E) + 2c_1c_2\beta_{12} + c_2^2(\alpha_2 - E) = 0$$

The coefficients c_1 and c_2 must be so chosen that $\mathbf{E}$ becomes minimum, *i.e.*,

$$\frac{\delta E}{\delta C_1} = 0 \text{ and } \frac{\delta E}{\delta C_2} = 0$$

The condition, $\frac{\delta E}{\delta C_1} = 0$ gives

$$c_1(\alpha_1 - E) + c_2\beta_{12} = 0 \tag{11.21}$$

Similarly, the condition, $\frac{\delta E}{\delta C_2} = 0$ gives

$$c_1\beta_{12} + c_2(\alpha_2 - E) = 0 \tag{11.22}$$

In case of Ethylene, $\alpha_1 = \alpha_2 = \alpha$ and $\beta_{12} = \beta_{21} = \beta$. Then the two linear equations give non-trivial solutions for C_1 and C_2 only if the following determinant holds:

$$\begin{vmatrix} \alpha - E & \beta \\ \beta & \alpha - E \end{vmatrix} = 0 \tag{11.23}$$

Now if β is zero, the Eqn. gives $(\alpha - E)^2 = 0$, *i.e.*, , $E = \alpha$. Thus, if the exchange of electrons between the two C-atoms in ethylene was absent, the energy of the π-electron would be same as that if it was localized on a particular C atom. This is entropically unfavourable for bonding to occur, since,formation of molecules from atom leads to decrease in entropy. Hence, for free energy change to be negative, energy must be released in the process. Therefore, for π-bonding to occur, the energy of a π-electron must be different from α; thus β can not be zero (that is, exchange of π-electrons between adjacent atoms must be allowed). The sign of β is obtained from the following reasoning. Whatever be the sign of β, if $\beta \neq 0$, we have from Eq. 11.23,

$$(\alpha - E)^2 = \beta^2 \;\; i.e., \; E = \alpha \pm \beta$$

Thus, we get two π-energy levels different from α. Now, we have to evaluate the value of c_1 and c_2.

For $E = \alpha + \beta$, from Eq. 11.23 we get,

$$c_1(\alpha - \alpha - \beta) + c_2\beta = 0 \text{ or, } c_1 = c_2$$

$$c_1\beta + c_2(\alpha - \alpha - \beta) = 0 \text{ or, } c_1 = c_2$$

For normalization of the wave function, $c_1^2 + c_2^2 = 1$, *i.e.*, $c_1 = c_2 = 1/\sqrt{2}$

Thus, the π-MO wave function corresponding to $E = \alpha + \beta$ would be:

$$\psi_{MO} = 1/\sqrt{2}(\phi_1 + \phi_2)$$

Since, this is an *in-phase* combination of the two atomic orbitals, ϕ_1, ϕ_2, the resulting MO must be of *bonding* in nature and its energy, $\alpha + \beta$, should be less than α according to the entropy based argument given earlier. Hence, β is negative.

The *antibonding* MO wave function corresponding to energy $E = \alpha - \beta$ can be similarly found to be:

$$\psi_{MO} = 1/\sqrt{2}(\phi_1 - \phi_2)$$

So the energy level diagram for the bonding and *antibonding* MOs of the π-electrons of ethylene are as shown below in Figure 11.15:

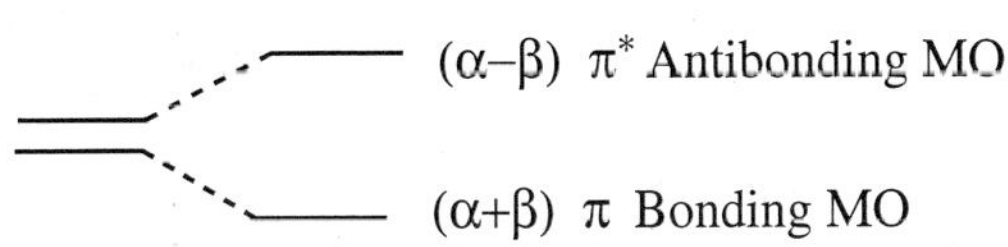

Figure 11.15: MOs of ethylene molecule.

Here one point should be noted that *bonding* and *antibonding* MOs are equally lowered and raised respectively, as here S_{ij} has been assumed to be zero.

Thus, so far we did not require the help of group theory to obtain the wave functions and their corresponding energies. Now at this stage, introduction of group theory is required to examine whether π (singlet) to π^* (singlet) electronic transition is allowed or not.

The point group of **Ethylene** is D_{2h}, where the molecule is in xy-plane. The atomic orbitals, ϕ_1 and ϕ_2 are perpendicular to the xy-plane and their positive lobes are directed towards the positive z-direction (Figure 11.16).We will try to see the effect of different symmetry operations on the two constituent atomic orbitals, ϕ_1 and ϕ_2 for bonding π-MO wave function.

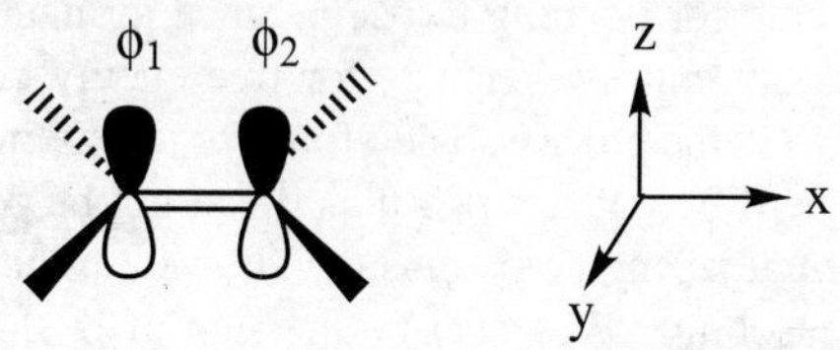

Figure 11.16: Two p_z orbitals of two C-atoms of ethylene molecule as basis.

$$E[\frac{1}{\sqrt{2}}(\phi_1+\phi_2)] = 1.[\frac{1}{\sqrt{2}}(\phi_1+\phi_2)]; \text{Hence, character for } E,\ \chi_{[E]} = 1$$

$$C_2(z)[\frac{1}{\sqrt{2}}(\phi_1+\phi_2)] = 1.[\frac{1}{\sqrt{2}}(\phi_1+\phi_2)]; \text{ Hence, character for } C_2(z),\ \chi_{[C_2(z)]} = 1$$

$$C_2(y)[\frac{1}{\sqrt{2}}(\phi_1+\phi_2)] = -1.[\frac{1}{\sqrt{2}}(-\phi_1-\phi_2)]; \text{ Hence, character for } C_2(y),\ \chi_{[C_2(y)]} = -1$$

$$C_2(x)[\frac{1}{\sqrt{2}}(\phi_1+\phi_2)] = -1.[\frac{1}{\sqrt{2}}(-\phi_1-\phi_2)]; \text{ Hence, character for } C_2(x),\ \chi_{[C_2(x)]} = -1$$

$$i[\frac{1}{\sqrt{2}}(\phi_1+\phi_2)] = -1.[\frac{1}{\sqrt{2}}(-\phi_1-\phi_2)]; \text{ Hence, character for } i,\ \chi_{[i]} = -1$$

$$\sigma(xy)[\frac{1}{\sqrt{2}}(\phi_1+\phi_2)] = -1.[\frac{1}{\sqrt{2}}(-\phi_1-\phi_2)]; \text{ Hence, character for } \sigma(xy),\ \chi_{[\sigma(xy)]} = -1$$

$$\sigma(xz)[\frac{1}{\sqrt{2}}(\phi_1+\phi_2)] = 1.[\frac{1}{\sqrt{2}}(\phi_1+\phi_2)]; \text{ Hence, character for } \sigma(xz),\ \chi_{[\sigma(xz)]} = 1$$

$$\sigma(yz)[\frac{1}{\sqrt{2}}(\phi_1+\phi_2)] = 1.[\frac{1}{\sqrt{2}}(\phi_1+\phi_2)]; \text{ Hence, character for } \sigma(yz),\ \chi_{[\sigma(yz)]} = 1$$

Here, we see that under some symmetry operations, both ϕ_1 and ϕ_2 remain unaltered and under some symmetry operations, both ϕ_1 and ϕ_2 changes their sign and accordingly we get +1 or –1 as coefficient. We can write the characters in a tabular form as follows:

Table 11.13: Characters corresponding to π-MO for ethylene molecule.

$\chi(E)$	$\chi(C_2(z))$	$\chi(C_2(y))$	$\chi(C_2(x))$	$\chi(i)$	$\chi(\sigma(xy))$	$\chi(\sigma(xz))$	$\chi(\sigma(yz))$
1	1	–1	–1	–1	–1	1	1

The above table (Table 11.13) corresponds to the irreducible representation B_{1u} in the character table of D_{2h}. Thus, the bonding π-MO wave function of ethylene acts as the basis for B_{1u} irreducible representation.

Similarly, we can generate the character for the π^*-MO by considering two p_z orbitals in opposite phase and see the effect of all the symmetry operations, present in the point group, on two functions ϕ_1 and ϕ_2 as:

$$E[\frac{1}{\sqrt{2}}(\phi_1-\phi_2)] = \frac{1}{\sqrt{2}}(\phi_1-\phi_2) = 1.[\frac{1}{\sqrt{2}}(\phi_1-\phi_2)];\ \text{Hence},\ \chi_{[E]} = 1$$

$$C_2(z)[\frac{1}{\sqrt{2}}(\phi_1-\phi_2)] = \frac{1}{\sqrt{2}}(\phi_2-\phi_1) = -1.[\frac{1}{\sqrt{2}}(\phi_1-\phi_2)];\ \text{Hence},\ \chi_{[C_2(z)]} = -1$$

$$C_2(y)[\frac{1}{\sqrt{2}}(\phi_1-\phi_2)] = \frac{1}{\sqrt{2}}(-\phi_2-(-\phi_1)) = 1.[\frac{1}{\sqrt{2}}(\phi_1-\phi_2)];\ \text{Hence},\ \chi_{[C_2(y)]} = 1$$

$$C_2(x)[\frac{1}{\sqrt{2}}(\phi_1-\phi_2)] = \frac{1}{\sqrt{2}}(-\phi_1-(-\phi_2)) = -1.[\frac{1}{\sqrt{2}}(\phi_1-\phi_2)];\ \text{Hence},\ \chi_{[C_2(x)]} = -1$$

$$i[\frac{1}{\sqrt{2}}(\phi_1-\phi_2)] = \frac{1}{\sqrt{2}}(-\phi_2-(-\phi_1)) = 1.[\frac{1}{\sqrt{2}}(\phi_1-\phi_2)];\ \text{Hence},\ \chi_{[i]} = 1$$

$$\sigma(xy)[\frac{1}{\sqrt{2}}(\phi_1-\phi_2)] = \frac{1}{\sqrt{2}}(-\phi_1-(-\phi_2)) = -1.[\frac{1}{\sqrt{2}}(\phi_1-\phi_2)];\ \text{Hence},\ \chi_{[\sigma(xy)]} = -1$$

$$\sigma(xz)[\frac{1}{\sqrt{2}}(\phi_1-\phi_2)] = \frac{1}{\sqrt{2}}(\phi_1-\phi_2) = 1.[\frac{1}{\sqrt{2}}(\phi_1-\phi_2)];\ \text{Hence},\ \chi_{[\sigma(xz)]} = 1$$

$$\sigma(yz)[\frac{1}{\sqrt{2}}(\phi_1-\phi_2)] = \frac{1}{\sqrt{2}}(\phi_2-\phi_1) = -1.[\frac{1}{\sqrt{2}}(\phi_1-\phi_2)];\ \text{Hence},\ \chi_{[\sigma(yz)]} = -1$$

After tabulation of the characters, thus obtained, we get,

Table 11.14: Characters corresponding to π^* MO for ethylene molecule.

$\chi(E)$	$\chi(C_2(z))$	$\chi(C_2(y))$	$\chi(C_2(x))$	$\chi(i)$	$\chi(\sigma(xy))$	$\chi(\sigma(xz))$	$\chi(\sigma(yz))$
1	–1	1	–1	1	–1	1	–1

The characters are found to be corroborated with B_{2g} IR of the character table of D_{2h}.

Now, alternatively,using projection operation method, we will try to find the linear combination form of the bonding and anti bonding MOs.

Let us consider that the ethylene molecule is in xy-plane and ϕ_1 and ϕ_2 corresponding to two p_z orbitals of two C-atoms, are on the plane of the paper (Figure 11.16). The following table (Table 11.15) shows the effect of different symmetry operations of D_{2h} point group on these two bases, ϕ_1 and ϕ_2 which are used as generating functions for projection operation method.

Table 11.15: Transformation table for D_{2h} symmetry group taking ϕ_1 and ϕ_2 as generating function.

D_{2h}	E	$C_2(z)$	$C_2(y)$	$C_2(x)$	i	$\sigma(xy)$	$\sigma(xz)$	$\sigma(yz)$
ϕ_1	ϕ_1	ϕ_2	$-\phi_2$	$-\phi_1$	$-\phi_2$	$-\phi_1$	ϕ_1	ϕ_2
ϕ_2	ϕ_2	ϕ_1	$-\phi_1$	$-\phi_2$	$-\phi_1$	$-\phi_2$	ϕ_2	ϕ_1

For A_g IR, taking ϕ_1 as generating function, projection operation results, ${}^{A_g}P_{(\phi_1)} = 0$. Similarly, for other irreducible representations

$${}^{B_{1g}}P_{(\phi_1)} = 0;\quad {}^{B_{2g}}P_{(\phi_1)} = 4(\phi_1-\phi_2);\quad {}^{B_{3g}}P_{(\phi_1)} = 0;\quad {}^{A_u}P_{(\phi_1)} = 0;$$

$$^{B_{1u}}P_{(\phi_1)} = 4(\phi_1 + \phi_2); \quad ^{B_{2u}}P_{(\phi_1)} = 0; \quad ^{B_{3u}}P_{(\phi_1)} = 0$$

Therefore, two linear combinations are possible with those two atomic orbitals;

$$B_{1u} \text{ symmetry: } \frac{1}{\sqrt{2}}(\phi_1 + \phi_2) \text{ and } B_{2g} \text{ symmetry: } \frac{1}{\sqrt{2}}(\phi_1 - \phi_2)$$

B_{1u} symmetry SALC is said to be of lower energy as it has constructive combination and B_{2g} symmetry SALC is said to be of higher energy as it has destructive combination . The energy levels are shown in Figure 11.17. It is apparent that the ground

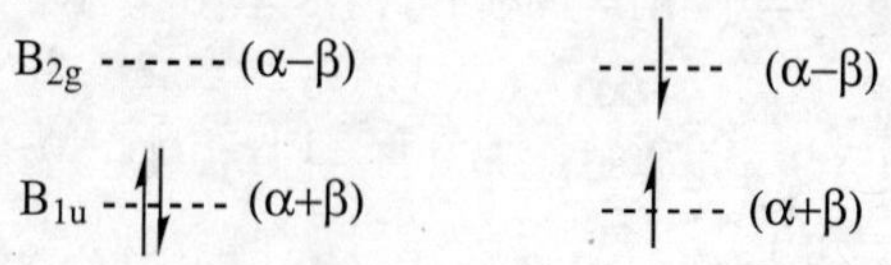

Ground state (singlet) Excited state (singlet)

Figure 11.17: Two energy states of ethylene molecule.

state electronic configuration of ethylene is $(b_{1u})^2$. Hence, the symmetry of the overall ground state configuration is the direct product of B_{1u} with itself and obviously it is going to result a totally symmetric representation, *i.e.*,

$$B_{1u} \times B_{1u} = A_g$$

In its singlet excited state, the electronic configuration is $(b_{1u})^1(b_{2g})^1$. The symmetry of the state corresponding to this electronic configuration would be:

$$(b_{1u}) \times (b_{2g}) = B_{3u}$$

Therefore, in terms of symmetry, in ethylene, the π (singlet) to π* (singlet) electronic transition is basically a transition from A_g symmetry to B_{3u} symmetry.

It is well known that the intensity or probability of any transition is proportional to the square of the transition moment integral, $M = \langle \psi_{gr}|\mu|\psi_{ex}\rangle$. Since, the symmetry of the ground and excited states are just determined and the symmetry of the three components of μ (μ_x, μ_y and μ_z correspond to B_{3u}, B_{2u} and B_{1u}, respectively) are evident from the D_{2h} character table, hence, it has now become a very easy task to find the transition moment integrals.

$$M_x = \langle \psi_{gr}|\mu_x|\psi_{ex}\rangle = A_g \times B_{3u} \times B_{3u} = A_g;$$

Hence, $M_x \neq 0$ *i.e.*, this component survives.

$$M_y = \langle \psi_{gr}|\mu_y|\psi_{ex}\rangle = A_g \times B_{2u} \times B_{3u} \neq A_g;$$

Hence, $M_y = 0$ *i.e.*, this component vanishes.

$$M_z = \langle \psi_{gr}|\mu_z|\psi_{ex}\rangle = A_g \times B_{1u} \times B_{3u} \neq A_g;$$

Hence, $M_z = 0$ *i.e.*, this component vanishes.

Thus, the transition moment integral survives because of the M_x component and hence the transition intensity (I) is non-zero. Therefore, it can be concluded that the transition is allowed and also polarized in the x-direction. This means that as a result of the transition, the electronic charge distribution in the molecule changes in such a way that only the x-component of the dipole moment undergoes major changes, while the changes in y and z-components are insignificant.

11.3.2 *cis*- and *trans*- 1,3 Butadiene

Let us first take the example of ***trans*-isomer** (C_{2h}) for discussion.

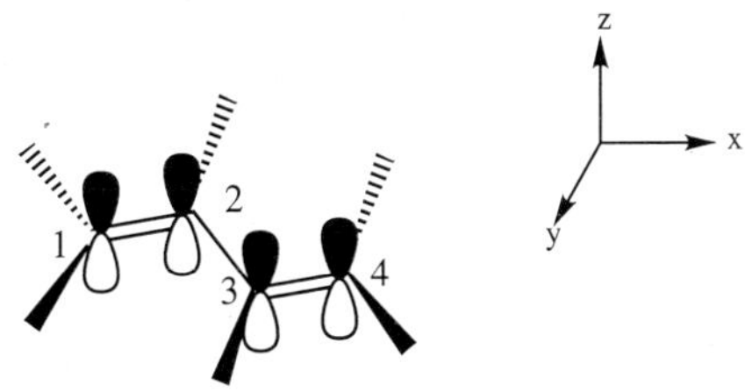

Figure 11.18: *trans* 1,3 butadiene molecule along with its four p_z orbitals.

Here, C_2 axis is perpendicular to the molecular plane (xy), *i.e.*, along z-axis. Consider the four p_z orbitals as basis (ϕ_i) and see how they transform under the different operations of C_{2h} point group (shown in Table 11.16). The reducible representation has also been presented in the same table.

Table 11.16: Transformation table for C_{2h} taking ϕ_1, ϕ_2, ϕ_3 and ϕ_4 as generating functions along with the reducible representation.

C_{2h}	E	$C_2(z)$	i	σ_h
ϕ_1	ϕ_1	ϕ_4	$-\phi_4$	$-\phi_1$
ϕ_2	ϕ_2	ϕ_3	$-\phi_3$	$-\phi_2$
ϕ_3	ϕ_3	ϕ_2	$-\phi_2$	$-\phi_3$
ϕ_4	ϕ_4	ϕ_1	$-\phi_1$	$-\phi_4$
Γ_{red}	4	0	0	–4

Using reduction formula we get,

$$\Gamma_{red} = 2A_u + 2B_g$$

To get the actual form and extent of participation of atomic orbitals in the formation of molecular orbitals, we need to take the help of projection operator technique.

- Let us take the projection operator method taking ϕ_1 as generating function.

$$\text{For IR } A_g, \sum[\chi_R(A_g).R\phi_1] = [1.\phi_1 + 1.\phi_4 + 1.(-\phi_4) + 1.(-\phi_1)] = 0$$

$$\text{For IR } B_g, \sum[\chi_R(B_g).R\phi_1] = [1.\phi_1 + (-1).\phi_4 + 1.(-\phi_4) + (-1).(-\phi_1)]$$
$$= 2(\phi_1 - \phi_4)$$

$$\text{For IR } A_u, \sum[\chi_R(A_u).R\phi_1] = [1.\phi_1 + 1.\phi_4 + (-1).(-\phi_4) + (-1).(-\phi_1)]$$
$$= 2(\phi_1 + \phi_4)$$

$$\text{For IR } B_u, \sum[\chi_R(B_u).R\phi_1] = [1.\phi_1 + (-1).\phi_4 + (-1).(-\phi_4) + 1.(-\phi_1)] = 0$$

- Taking ϕ_2 as generating function.

$$\text{For IR } A_g, \sum[\chi_R(A_g).R\phi_2] = [1.\phi_2 + 1.\phi_3 + 1.(-\phi_3) + 1.(-\phi_2)] = 0$$

$$\text{For IR } B_g, \sum[\chi_R(B_g).R\phi_2] = [1.\phi_2 + (-1).\phi_3 + 1.(-\phi_3) + (-1).(-\phi_2)] = 2(\phi_2 - \phi_3)$$

$$\text{For IR } A_u, \sum[\chi_R(A_u).R\phi_2] = [1.\phi_2 + 1.\phi_3 + (-1).(-\phi_3) + (-1).(-\phi_2)] = 2(\phi_2 + \phi_3)$$

$$\text{For IR } B_u, \sum[\chi_R(B_u).R\phi_2] = [1.\phi_2 + 1.\phi_3 + (-1).(-\phi_3) + (-1).(-\phi_2)] = 0$$

- Taking ϕ_3 as generating function.

$$\text{For IR } A_g, \sum[\chi_R(A_g).R\phi_3] = [1.\phi_3 + 1.\phi_2 + 1.(-\phi_2) + 1.(-\phi_3)] = 0$$

$$\text{For IR } B_g, \sum[\chi_R(B_g).R\phi_3] = [1.\phi_3 + (-1).\phi_2 + 1.(-\phi_2) + (-1).(-\phi_3)] = -2(\phi_2 - \phi_3)$$

$$\text{For IR } A_u, \sum[\chi_R(A_u).R\phi_3] = [1.\phi_3 + 1.\phi_2 + (-1).(-\phi_2) + (-1).(-\phi_3)] = 2(\phi_2 + \phi_3)$$

$$\text{For IR } B_u, \sum[\chi_R(B_u).R\phi_3] = [1.\phi_3 + (-1).\phi_2 + (-1).(-\phi_2) + 1.(-\phi_3)] = 0$$

- Taking ϕ_4 as generating function.

$$\text{For IR } A_g, \sum[\chi_R(A_g).R\phi_4] = [1.\phi_4 + 1.\phi_1 + 1.(-\phi_1) + 1.(-\phi_4)] = 0$$

$$\text{For IR } B_g, \sum[\chi_R(B_g).R\phi_4] = [1.\phi_4 + (-1).\phi_1 + 1.(-\phi_1) + (-1).(-\phi_4)] = -2(\phi_1 - \phi_4)$$

$$\text{For IR } A_u, \sum[\chi_R(A_u).R\phi_4] = [1.\phi_4 + 1.\phi_1 + (-1).(-\phi_1) + (-1).(-\phi_4)] = 2(\phi_1 + \phi_4)$$

$$\text{For IR } B_u, \sum[\chi_R(B_u).R\phi_4] = [1.\phi_4 + (-1).\phi_1 + (-1).(-\phi_1) + 1.(-\phi_4)] = 0$$

We see that corresponding to A_u symmetry, linear combination, $(\phi_1 + \phi_4)$ is obtained when ϕ_1 and ϕ_4 are used as generating function, whereas, linear combination,$(\phi_2 + \phi_3)$ is obtained when ϕ_2 and ϕ_3 are used as generating function.

Similarly, corresponding to B_g symmetry two linear combinations, $(\phi_1 - \phi_4)$ and $(\phi_2 - \phi_3)$ are obtained when ϕ_1, ϕ_4 and ϕ_2, ϕ_3 respectively, are used as generating function.

Therefore, the two positive combinations are of A_u symmetry and the two negative combinations are of B_g symmetry. For Butadiene, by solving the wave functions, the following linear combinations for the four MOs are:

A_u symmetry:

$$\Phi_1 = 0.37(\phi_1 + \phi_4) + 0.60(\phi_2 + \phi_3); \qquad \Phi_3 = 0.60(\phi_1 + \phi_4) - 0.37(\phi_2 + \phi_3)$$

B_g symmetry:

$$\Phi_2 = 0.60(\phi_1 - \phi_4) + 0.37(\phi_2 - \phi_3); \qquad \Phi_4 = 0.37(\phi_1 - \phi_4) - 0.60(\phi_2 - \phi_3)$$

The magnitude of coefficients depicts the extent of contribution of the corresponding wave function in total molecular orbital. Pictorially, the MOs along with their relative energies are shown in Figure 11.19. As the number of nodes increases, the relative energy increases as well. The orientation and the size of the lobes describe the nature of combination (whether in phase or out of phase) and its contribution, respectively.

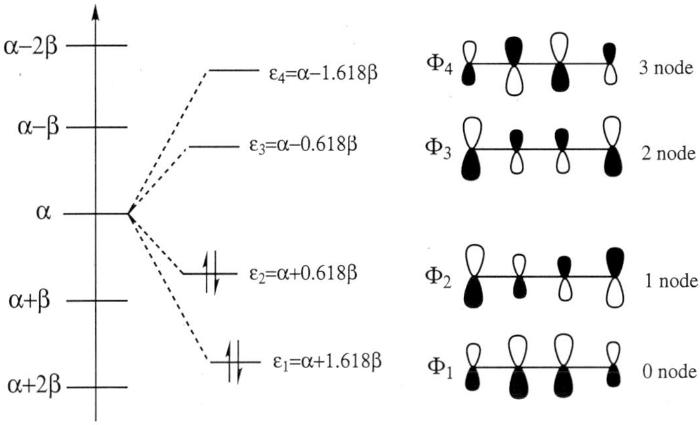

Figure 11.19: MO diagram of *trans* 1,3 butadiene.

The ground state configuration of *trans*-butadiene is $(\Phi_1)^2(\Phi_2)^2$, which is of singlet A_g symmetry, *i.e.*, ${}^1A_g$. The first excited state configuration is $(\Phi_1)^2(\Phi_2)^1(\Phi_3)^1$, or, it can be written as, $(a_u)^2(b_g)^1(a_u)^1$. Hence, the overall symmetry of this excited state would be: $A_g \times B_g \times A_u = B_u$. This resulting state may be of singlet, ${}^1B_u$ or triplet ${}^3B_u$ type. Now, let us calculate the transition moment integral for the transition between two singlet states, ${}^1A_g$ and ${}^1B_u$. The symmetry of the three components of μ (μ_x, μ_y and μ_z correspond to B_u, B_u and A_u, respectively

$$M_x = \langle \psi_{gr} | \mu_x | \psi_{ex} \rangle = A_g \times B_u \times B_u = A_g;$$

Hence, $M_x \neq 0$ *i.e.*, this component survives.

$$M_y = \langle \psi_{gr} | \mu_y | \psi_{ex} \rangle = A_g \times B_u \times B_u = A_g;$$

Hence, $M_y \neq 0$ *i.e.*, this component survives.

$$M_z = \langle \psi_{gr} | \mu_z | \psi_{ex} \rangle = A_g \times A_u \times B_u = B_g;$$

Hence, $M_z = 0$ *i.e.*, this component vanishes.
Therefore, this transition is allowed and found to be *xy*-polarised. Although, transition between ${}^1A_g$ and ${}^3B_u$ states is, in a similar way, are found to be orbital allowed but spin forbidden. Since the later transition is between states of two dissimilar multiplicities.

We can continue the discussion with ***cis*-1,3 butadiene** (C_{2v} point group) molecule as well.

Consider that the molecule is in *xy* plane and C_2 axis is along *y*-direction *i.e.*, perpendicular to plane of the paper. Hence, π- electrons are in $2p_z$ orbitals and taken

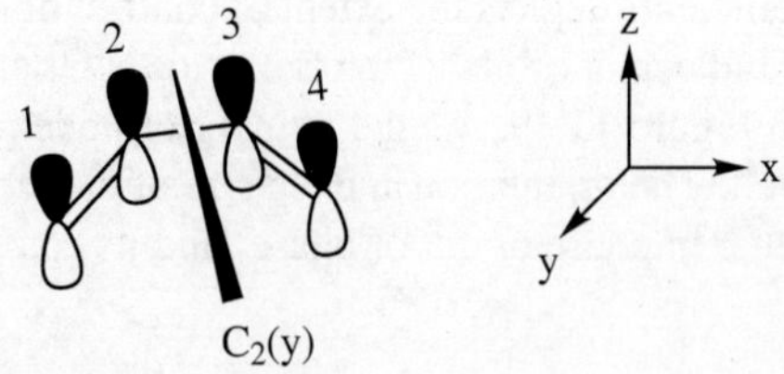

Figure 11.20: *cis* 1,3 butadiene molecule along with its four p_z orbitals.

as basis and designated as ϕ_i. We shall see how they transform under various operations of C_{2v} point group and presented in Table 11.17. The reducible representation is also presented in the last row of the table.

Table 11.17: Transformation table for C_{2v} taking ϕ_1, ϕ_2, ϕ_3 and ϕ_4 as generating functions along with the reducible representation presented in the last row.

C_{2v}	E	$C_2(y)$	σ_{xz}	σ_{yz}
ϕ_1	ϕ_1	ϕ_4	$-\phi_4$	$-\phi_1$
ϕ_2	ϕ_2	ϕ_3	$-\phi_3$	$-\phi_2$
ϕ_3	ϕ_3	ϕ_2	$-\phi_2$	$-\phi_3$
ϕ_4	ϕ_4	ϕ_1	$-\phi_1$	$-\phi_4$
Γ_{red}	4	0	0	–4

After reduction, we get,

$$\Gamma_{\text{red}} = 2A_2 + 2B_1$$

Although, it is evident that there should not be any linear combination corresponding to the IRs, A_1 and B_2 symmetry, still calculations are shown for those two IRs.

- Let us take the projection operator method taking ϕ_1 as generating function.

$$\text{For IR } A_1, \sum[\chi_R(A_1).R\phi_1] = [1.\phi_1 + 1.(-\phi_4) + 1.(-\phi_1) + 1.(\phi_4)] = 0$$

$$\text{For IR } A_2, \sum[\chi_R(A_2).R\phi_1] = [1.\phi_1 + 1.(-\phi_4) + (-1).(-\phi_1) + (-1).(\phi_4)]$$

$$= 2(\phi_1 - \phi_4)$$

$$\text{For IR } B_1, \sum[\chi_R(B_1).R\phi_1] = [1.\phi_1 + (-1).(-\phi_4) + (-1).(-\phi_1) + 1.\phi_1]$$

$$= 2(\phi_1 + \phi_4)$$

$$\text{For IR } B_2, \sum[\chi_R(B_2).R\phi_1] = [1.\phi_1 + (-1).(-\phi_4) + 1.(-\phi_1) + (-1).(-\phi_4)] = 0$$

- Taking ϕ_2 as generating function.

$$\text{For IR } A_1, \sum[\chi_R(A_1).R\phi_2] = [1.\phi_2 + 1.(-\phi_3) + 1.(-\phi_2) + 1.\phi_3] = 0$$

$$\text{For IR } A_2, \sum[\chi_R(A_2).R\phi_2] = [1.\phi_2 + 1.(-\phi_3) + (-1).(-\phi_2) + (-1).\phi_3)]$$

$$= 2(\phi_2 - \phi_3)$$

$$\text{For IR } B_1, \sum[\chi_R(B_1).R\phi_2] = [1.\phi_2 + (-1).(-\phi_3) + (-1).(-\phi_2) + 1.\phi_3]$$
$$= 2(\phi_2 + \phi_3)$$
$$\text{For IR } B_2, \sum[\chi_R(B_2).R\phi_2] = [1.\phi_2 + (-1).(-\phi_3) + 1.(-\phi_2) + (-1).\phi_3] = 0$$

- Taking ϕ_3 as generating function.

$$\text{For IR } A_1, \sum[\chi_R(A_1).R\phi_3] = [1.\phi_3 + 1.(-\phi_2) + 1.(-\phi_3) + 1.\phi_2] = 0$$
$$\text{For IR } A_2, \sum[\chi_R(A_2).R\phi_3] = [1.\phi_3 + 1.(-\phi_2) + (-1).(-\phi_3) + (-1).\phi_2]$$
$$= -2(\phi_2 - \phi_3)$$
$$\text{For IR } B_1, \sum[\chi_R(B_1).R\phi_3] = [1.\phi_3 + (-1).(-\phi_2) + (-1).(-\phi_3) + 1.\phi_2]$$
$$= 2(\phi_2 + \phi_3)$$
$$\text{For IR } B_2, \sum[\chi_R(B_2).R\phi_3] = [1.\phi_3 + (-1).(-\phi_2) + 1.(-\phi_3) + (-1).\phi_2] = 0$$

- Taking ϕ_4 as generating function.

$$\text{For IR } A_1, \sum[\chi_R(A_1).R\phi_4] = [1.\phi_1 + 1.(-\phi_4) + 1.(-\phi_1) + 1.\phi_4] = 0$$
$$\text{For IR } A_2, \sum[\chi_R(A_2).R\phi_4] = [1.\phi_1 + 1.(-\phi_1) + (-1).(-\phi_1) + (-1).\phi_4]$$
$$= 2(\phi_1 - \phi_4)$$
$$\text{For IR } B_1, \sum[\chi_R(B_1).R\phi_4] = [1.\phi_1 + (-1).(-\phi_4) + (-1).(-\phi_1) + 1.\phi_4]$$
$$= 2(\phi_1 + \phi_4)$$
$$\text{For IR } B_2, \sum[\chi_R(B_2).R\phi_4] = [1.\phi_1 + (-1).(-\phi_4) + 1.(-\phi_1) + (-1).\phi_4] = 0$$

It is found that out of the four different symmetries, only A_2 adopts the two out-of-phase linear combination of $(\phi_1 - \phi_4)$ and $(\phi_2 - \phi_3)$ and B_1 adopts the two in-phase linear combination of $(\phi_1 + \phi_4)$ and $(\phi_2 + \phi_3)$. We can recall the equations of MOs for butadiene,

$$\Phi_1 = 0.37(\phi_1 + \phi_4) + 0.60(\phi_2 + \phi_3); \quad \Phi_3 = 0.60(\phi_1 + \phi_4) - 0.37(\phi_2 + \phi_3)$$
$$\Phi_2 = 0.60(\phi_1 - \phi_4) + 0.37(\phi_2 - \phi_3); \quad \Phi_4 = 0.37(\phi_1 - \phi_4) - 0.60(\phi_2 - \phi_3)$$

Here, we can see that Φ_1 and Φ_3 are of B_1 symmetry and Φ_2 and Φ_4 are of A_2 symmetry.

The ground state configuration of *cis*-butadiene is $(\Phi_1)^2(\Phi_2)^2$, which is of singlet A_1 symmetry, *i.e.*, $^1A_1$. The first excited state configuration is $(\Phi_1)^2(\Phi_2)^1(\Phi_3)^1$, or, it can be written as, $(b_1)^2(a_2)^1(b_1)^1$. Hence, the overall symmetry of this excited state would be. $A_1 \times A_2 \times B_1 = B_2$. This resulting state may be of singlet, $^1B_2$ or triplet $^3B_2$ type. Now, let us calculate the transition moment integral for the transition between two singlet states, $^1A_1$ and $^1B_2$. The symmetry of the three components of μ (μ_x, μ_y and μ_z correspond to B_1, B_2 and A_1, respectively.

$$M_x = \langle \psi_{gr} | \mu_x | \psi_{ex} \rangle = A_1 \times B_1 \times B_2 = A_2;$$

Hence, $M_x = 0$ *i.e.*, this component vanishes.

$$M_y = \langle \psi_{gr} | \mu_y | \psi_{ex} \rangle = A_1 \times B_2 \times B_2 = A_1;$$

Hence, $M_y \neq 0$ *i.e.*, this component survives.

$$M_z = \langle \psi_{gr} | \mu_z | \psi_{ex} \rangle = A_1 \times A_1 \times B_2 = B_2;$$

Hence, $M_z = 0$ *i.e.*, this component vanishes.
Therefore, this transition (Φ_2 to Φ_3) is allowed due to the presence of y- component of dipole momemnt and hence y-polarised. In a similar way other possible transitions (Φ_2 to Φ_4 or Φ_1 to Φ_3) can also be calculated.

11.3.3 Benzene (D_{6h})

Benzene molecule is on xy-plane and p_z orbitals for each C-atoms are along z-axis. Each C-atom undergoes sp^2 hybridization for the cyclic σ-bond framework, involving s and p_x, p_y orbitals. Here, only we need to consider the interactions among those six p_z atomic orbitals ($\phi_1, \phi_2, \phi_3, \phi_4, \phi_5$ and ϕ_6, shown in Figure 11.21), which remain untouched during σ-bondings, combine to form six π-MO's. Taking those six π-AO's along with their signs of lobes is considered as basis for construction of reducible representation shown in Table 11.18.

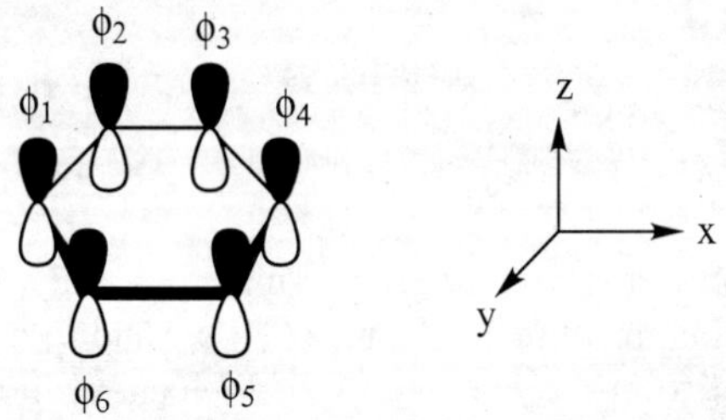

Figure 11.21: Numbering system of P_z orbitals of benzene molecule along with the coordinate system.

Table 11.18: Reducible representation for D_{6h} point group, taking six ϕ_i's as basis for benzene molecule.

D_{6h}	E	$2C_6$	$2C_3$	C_2	$3C_2'$	$3C_2''$	i	$2S_3$	$2S_6$	σ_h	$3\sigma_d$	$3\sigma_v$
Γ_π	6	0	0	0	–2	0	0	0	0	–6	0	2

On reduction to its irreducible components, we get,

$$\Gamma_\pi = B_{2g} + E_{1g} + A_{2u} + E_{2u}$$

It says that among the six MO's Two would be of nondegenerate B_{2g} and A_{2u} symmetries and other two pairs would be of doubly degenerate having E_{1g} and E_{2u} symmetry nature. Our next job would be to find the exact combinations of AO's during those MO formation. For this, the projection operation method would be employed and corresponding transformation table is also presented for our ready reference. Taking other ϕ's, as generating function, the table can further be extended.

Table 11.19: Transformation table for D_{6h} point group, taking ϕ_1 as generating function.

D_{6h}	E	C_6	C_6^5	C_3	C_3^2	C_2	$C_2'(1)$	$C_2'(2)$	$C_2'(3)$	$C_2''(1)$	$C_2''(2)$	$C_2''(3)$	i	S_3	S_3^2	S_6
ϕ_1	ϕ_1	ϕ_2	ϕ_6	ϕ_3	ϕ_5	ϕ_4	$-\phi_1$	$-\phi_3$	$-\phi_5$	$-\phi_2$	$-\phi_4$	$-\phi_6$	$-\phi_4$	$-\phi_3$	$-\phi_5$	$-\phi_2$

S_6^5	σ_h	$\sigma_d(1)$	$\sigma_d(2)$	$\sigma_d(3)$	$\sigma_v(1)$	$\sigma_v(2)$	$\sigma_v(3)$
$-\phi_6$	$-\phi_1$	ϕ_2	ϕ_4	ϕ_6	ϕ_1	ϕ_3	ϕ_5

Applying projection operation on A_{2u} symmetry to find the linear combination taking ϕ_1 as generating function.

- Linear combination corresponding to A_{2u} symmetry:

$$
\begin{aligned}
{}^{A_{2u}}P_{\phi_1} &= \sum[\chi_R(A_{2u}).R\phi_1] \\
&= [(1\times\phi_1)+(1\times\phi_2)+(1\times\phi_6)+(1\times\phi_3)+(1\times\phi_5)+(1\times\phi_4) \\
&\quad +(-1\times-\phi_1)+(-1\times-\phi_3)+(-1\times-\phi_5)+(-1\times-\phi_2) \\
&\quad +(-1\times-\phi_4)+(-1\times-\phi_6)+(-1\times-\phi_4) \\
&\quad +(-1\times-\phi_3)+(-1\times-\phi_5)+(-1\times-\phi_2)+(-1\times-\phi_6) \\
&\quad +(-1\times-\phi_1)+(1\times\phi_2)+(1\times\phi_4)+(1\times\phi_6)+(1\times\phi_1) \\
&\quad +(1\times\phi_3)+(1\times\phi_5)]
\end{aligned}
$$

$$= 4(\phi_1+\phi_2+\phi_3+\phi_4+\phi_5+\phi_6)$$

$$= \frac{1}{\sqrt{6}}(\phi_1+\phi_2+\phi_3+\phi_4+\phi_5+\phi_6);\ \text{After normalization.}$$

In a similar way, if any other ϕ's are taken as generating function, the same result is obtained every time. Hence, the nondegenerate SALC corresponding to A_{2u} symmetry is

$$\Omega_1 = \frac{1}{\sqrt{6}}(\phi_1+\phi_2+\phi_3+\phi_4+\phi_5+\phi_6)$$

- Linear combination corresponding to B_{2g} symmetry: Starting with ϕ_1 as generating function, in a similar way, the normalized form of the SALC would be:

$$\Omega_6 = \frac{1}{\sqrt{6}}(\phi_1-\phi_2+\phi_3-\phi_4+\phi_5-\phi_6)$$

It can be tried with other ϕ's, but equivalent SALC's are obtained in every case.

- Linear combinations corresponding to E_{1g} symmetry: To find the linear combinations of degenerate E_{1g} symmetry is not so straight forward like the earlier two cases. Already it has been shown in case of octahedral geometry (sec. 11.2.4) that to find the two partners of E_{1g} symmetry, we need to employ special technique after performing projection operation method. The two forms of the SALCs, thus obtained, has the following normalized forms:

$$\Omega_2 = \frac{1}{\sqrt{12}}(2\phi_1 + \phi_2 - \phi_3 + 2\phi_4 - \phi_5 - \phi_6)$$
$$\Omega_3 = \frac{1}{2}(\phi_2 + \phi_3 - \phi_5 - \phi_6)$$

- Linear combinations corresponding to E_{2u} symmetry: The two normalized SALCs are

$$\Omega_4 = \frac{1}{\sqrt{12}}(2\phi_1 - \phi_2 - \phi_3 + 2\phi_4 - \phi_5 - \phi_6)$$
$$\Omega_5 = \frac{1}{2}(\phi_2 - \phi_3 + \phi_5 - \phi_6)$$

The π-MOs, thus formed, are shown diagrammatically along with their nodal planes in Figure 11.22. The electronic transitions for benzene molecule following these MOs are discussed in sec. 9.2.4. Nothing to do with symmetry aspects during the calculation of energy levels corresponding to different π-MOs. Here, use of Huckel's approximation and secular determinant gives the energy values. The energy value of π_1 MO corresponding to the wave function, Ω_1 is $(\alpha + 2\beta)$ and is found to be the lowest one (α and β have their conventional meaning in quantum mechanics). There is a correlation between the energy of orbitals and number of nodal planes they possess. As the number of nodal planes increases, the energy increases as well. In case of π_1, there is no nodal plane, hence possess the least energy. Similarly, π_6 corresponding to the wave function, Ω_6 has the highest energy, $(\alpha - 2\beta)$ with a maximum three nodal planes. The two degenerate MOs, π_2 and π_3 (SALCs, Ω_2 and Ω_3 corresponding to E_{1g} symmetry have energy $(\alpha + \beta)$ with one nodal plane, whereas, other two degenerate MOs, π_4 and π_5 (SALCs, Ω_4 and Ω_5 corresponding to E_{2u} symmetry have energy $(\alpha - \beta)$ with two nodal planes. Therefore, according to the quantum chemical calculation as well as calculating the number of nodal planes present in wave functions corresponding to MOs, the energy levels can be arranged accordingly as shown in Figure 11.22.

At this point, two important parameters of benzene molecule can be calculated, namely, charge density on each C-atom and bond order of C-C bonds.

- **Charge density on C-atoms:**

  Charge density (q_r) on any C-atom can be calculated using the following equation:

$$q_r = \sum_i n_i |c_{ri}|^2$$

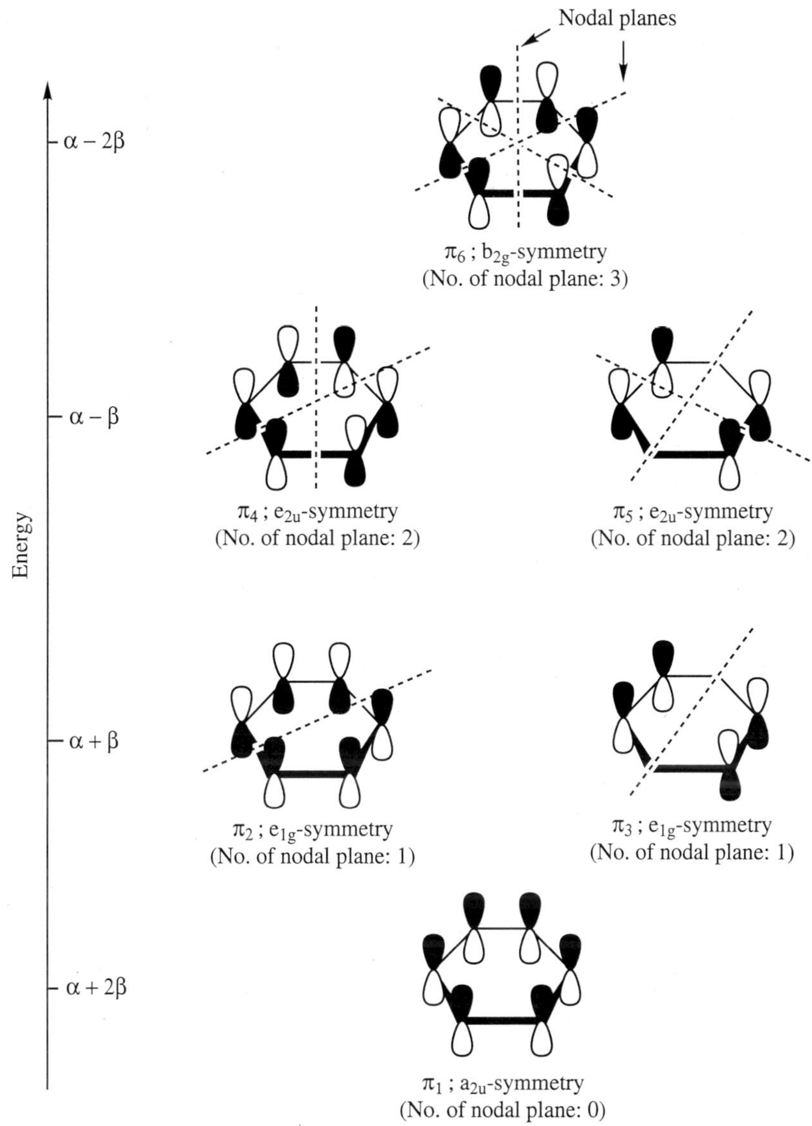

Figure 11.22: Pictorial representation of π MO's of benzene molecule.

where, n_i is the number of electrons in the ψ_i MO (which is known from electronic configuration) and c_{ri} is the coefficient of the $2p_z$ orbital of the ith C-atom of the rth π- MO.

For example, the charge density on C_1-atom,

$$[2\times(\frac{1}{\sqrt{6}})^2+2\times(\frac{2}{\sqrt{12}})^2+0+0\times(\frac{2}{\sqrt{12}})^2+0+0\times(\frac{1}{\sqrt{6}})^2]=1$$

Similarly, the charge density on C_3-atom,

$$[2\times(\frac{1}{\sqrt{6}})^2+2\times(\frac{1}{\sqrt{12}})^2+2\times(\frac{1}{2})^2+0\times(\frac{1}{\sqrt{12}})^2+0\times(\frac{1}{2})^2+0\times(\frac{1}{\sqrt{6}})^2]=1$$

In this way charge density on any C-atom can easily be calculated. It is found that all the C-atoms possess charge density of unity, *i.e.*, charges are equally distributed through out the carbon framework of benzene.

- **π-bond order of two adjacent C-atoms:**

  π-bond order B_{rs} between two adjacent C-atoms (C_r and C_s) can be calculated using the following equation:

$$B_{rs} = \sum_i n_i|c_{ri}||c_{si}|$$

where, c_{ri} and c_{si} are the coefficients of $2p_z$ orbitals of rth and s^{th} C-atoms in the ith π MO.

For example, bond order between C_1 and C_2 atoms,

$$\left[2\times\left(\frac{1}{\sqrt{6}}\right)\times\left(\frac{1}{\sqrt{6}}\right)+2\times\left(\frac{2}{\sqrt{12}}\right)\times\left(\frac{1}{\sqrt{12}}\right)+2\times(0)\times\left(\frac{1}{2}\right)+0\right.$$
$$\left.\times\left(\frac{2}{\sqrt{12}}\right)\times\left(\frac{1}{\sqrt{12}}\right)+0\times(0)\times\left(\frac{1}{2}\right)+0\times\left(\frac{1}{\sqrt{6}}\right)\times\left(\frac{1}{\sqrt{6}}\right)\right]=0.5$$

The same results are obtained for any two adjacent pair of benzene ring. This is because benzene has a delocalized π-electron system, and the six π electrons are equally shared among all six C-C bonds.

11.3.4 Naphthalene (D_{2h})

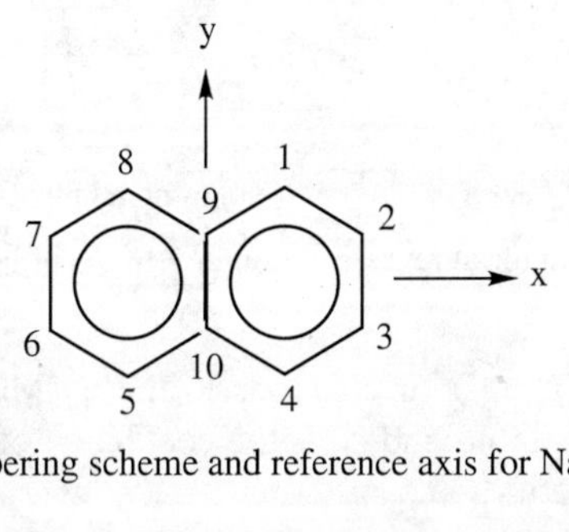

Figure 11.23: Numbering scheme and reference axis for Naphthalene molecule.

Let us consider the ten p_z orbitals of C-atoms, which take part in the formation of π-orbital system, as basis and are designated as ϕ_1 to ϕ_{10} (shown in Figure 11.23). The ten p_z orbitals of C-atoms are perpendicular to the molecular plane. The reducible representations can be determined based on the number of shifted and unshifted basis and then are tabulated as:

Table 11.20: Reducible representation for D_{2h} point group, taking ϕ_i's as basis.

D_{2h}	E	$C_2(z)$	$C_2(y)$	$C_2(x)$	i	σ_{xy}	σ_{xz}	σ_{yz}
Γ_π	10	0	0	–2	0	–10	2	0

After reduction we get,

$$\Gamma_\pi = 2B_{2g} + 3B_{3g} + 2A_u + 3B_{1u}$$

Hence, we see that two linear combinations for each B_{2g} and A_u symmetry and three linear combinations for each B_{3g} and B_{1u} symmetry. Now, we need to construct proper SALCs using projection operator corresponding to each IR component of Γ_π. At this moment from the structure of the molecule, one point must be noted that all the ϕ_i's are not equivalent. They can be grouped as:

Group 1: ϕ_1, ϕ_4,ϕ_5,ϕ_8

Group 2: ϕ_2,ϕ_3,ϕ_6,ϕ_7

Group 3: ϕ_9,ϕ_{10}

The transformation of any basis (ϕ's) will remain confined to that particular group. For example, if we try to see the effect of different symmetry operations of D_{2h} on ϕ_1 as generating function, after transformation, the result will always be within the members of *group 1* and similarly it is valid for other groups as well.

We can construct the transformation table and are presented in Table 11.21.

Table 11.21: Transformation table for D_{2h} point group, taking ϕ_i's as generating function.

D_{2h}	E	$C_2(z)$	$C_2(y)$	$C_2(x)$	i	σ_{xy}	σ_{xz}	σ_{yz}
ϕ_1	ϕ_1	ϕ_5	$-\phi_8$	$-\phi_4$	$-\phi_5$	$-\phi_1$	ϕ_4	ϕ_8
ϕ_2	ϕ_2	ϕ_6	$-\phi_7$	$-\phi_3$	$-\phi_6$	$-\phi_2$	ϕ_3	ϕ_7
ϕ_3	ϕ_3	ϕ_7	$-\phi_6$	$-\phi_2$	$-\phi_7$	$-\phi_3$	ϕ_2	ϕ_6
ϕ_4	ϕ_4	ϕ_8	$-\phi_5$	$-\phi_1$	$-\phi_8$	$-\phi_4$	ϕ_1	ϕ_5
ϕ_5	ϕ_5	ϕ_1	$-\phi_4$	$-\phi_8$	$-\phi_1$	$-\phi_5$	ϕ_8	ϕ_4
ϕ_6	ϕ_6	ϕ_2	$-\phi_3$	$-\phi_7$	$-\phi_2$	$-\phi_6$	ϕ_7	ϕ_3
ϕ_7	ϕ_7	ϕ_3	$-\phi_2$	$-\phi_6$	$-\phi_3$	$-\phi_7$	ϕ_6	ϕ_2
ϕ_8	ϕ_8	ϕ_4	$-\phi_1$	$-\phi_5$	$-\phi_4$	$-\phi_8$	ϕ_5	ϕ_1
ϕ_9	ϕ_9	ϕ_{10}	$-\phi_9$	$-\phi_{10}$	$-\phi_{10}$	$-\phi_9$	ϕ_{10}	ϕ_9
ϕ_{10}	ϕ_{10}	ϕ_9	$-\phi_{10}$	$-\phi_9$	$-\phi_9$	$-\phi_{10}$	ϕ_9	ϕ_{10}

Now, apply projection operation on A_u symmetry to extract the linear combination taking ϕ_1 as generating function.

- Two linear combinations of A_u symmetry:

$$^{A_u}P_{\phi_1} = \sum[\chi_R(A_u).R\phi_1]$$

$$= [1\times\phi_1 + 1\times\phi_5 + 1\times(-\phi_8) + 1\times(-\phi_4) + (-1)\times(-\phi_5) +$$

$$(-1)\times(-\phi_1) + (-1)\times\phi_4 + (-1)\times\phi_8]$$

$$= \phi_1 + \phi_5 - \phi_8 - \phi_4 + \phi_5 + \phi_1 - \phi_4 - \phi_8$$
$$= 2(\phi_1 - \phi_4 + \phi_5 - \phi_8)$$

Hence, after normalization, say,

$$\psi_1 = \frac{1}{2}(\phi_1 - \phi_4 + \phi_5 - \phi_8)$$

In a similar way, if projection operation is completed taking ϕ_4, ϕ_5 or ϕ_8 as generating function, exactly the same linear combination will be obtained since, they are all equivalent basis function or in the same group. To get other partner combination of A_u symmetry, we need to take generating function from other group. Let us try with ϕ_2 and find the combination.

$$^{A_u}P_{\phi_2} = \sum[\chi_R(A_u).R\phi_2]$$
$$= [1 \times \phi_2 + 1 \times \phi_6 + 1 \times (-\phi_7) + 1 \times (-\phi_3) + (-1) \times (-\phi_6) +$$
$$(-1) \times (-\phi_2) + (-1) \times \phi_3 + (-1) \times \phi_7]$$
$$= 2(\phi_2 - \phi_3 + \phi_6 - \phi_7)$$

Hence, after normalization, say,

$$\psi_2 = \frac{1}{2}(\phi_2 - \phi_3 + \phi_6 - \phi_7)$$

Since, already we have constructed two possible linear combinations of A_u symmetry, obviously no other combination is feasible. But still to verify this we will try projection operation taking other basis from the another group left, *i.e.*, ϕ_9 or ϕ_{10}. It is found that:

$$^{A_u}P_{\phi_9} = 0 \text{ and } ^{A_u}P_{\phi_{10}} = 0$$

- Three linear combinations of B_{1u} symmetry:

  In a similar way, taking three bases from the three groups, the three normalized SALCs for B_{1u} are:

$$\psi_3 = {}^{B_{1u}}P_{(\phi_1)} = \frac{1}{2}(\phi_1 + \phi_4 + \phi_5 + \phi_8)$$

$$\psi_4 = {}^{B_{1u}}P_{(\phi_2)} = \frac{1}{2}(\phi_2 + \phi_3 + \phi_6 + \phi_7)$$

$$\psi_5 = {}^{B_{1u}}P_{(\phi_9)} = \frac{1}{\sqrt{2}}(\phi_9 + \phi_{10})$$

- Two linear combinations of B_{2g} symmetry:

$$\psi_6 = {}^{B_{2g}}P_{(\phi_1)} = \frac{1}{2}(\phi_1 + \phi_4 - \phi_5 - \phi_8)$$

$$\psi_7 = {}^{B_{2g}}P_{(\phi_2)} = \frac{1}{2}(\phi_2 + \phi_3 - \phi_6 - \phi_7)$$

If projection operation method is tried taking either ϕ_9 or ϕ_{10} as generating function, result must be zero, *i.e.*, no further linear combination is possible for B_{2g} symmetry.

- Three linear combinations of B_{3g} symmetry:

$$\psi_8 = {}^{B_{3g}}P_{(\phi_1)} = \frac{1}{2}(\phi_1 - \phi_4 - \phi_5 + \phi_8)$$

$$\psi_9 = {}^{B_{3g}}P_{(\phi_2)} = \frac{1}{2}(\phi_2 - \phi_3 - \phi_6 + \phi_7)$$

$$\psi_{10} = {}^{B_{3g}}P_{(\phi_9)} = \frac{1}{\sqrt{2}}(\phi_9 - \phi_{10})$$

These ten SALCs, based on their similarity in symmetry, are further combined linearly to form π-MOs of naphthalene.

$$\Psi_1 = c_1\psi_1 + c_2\psi_2 \qquad (A_u \ symmetry)$$

$$\Psi_2 = c_3\psi_3 + c_4\psi_4 + c_5\psi_5 \qquad (B_{1u} \ symmetry)$$

$$\Psi_3 = c_6\psi_6 + c_7\psi_7 \qquad (B_{2g} \ symmetry)$$

$$\Psi_4 = c_8\psi_8 + c_9\psi_9 + c_{10}\psi_{10} \qquad (B_{3g} \ symmetry)$$

Now, Huckel approximation is applied to find the magnitude of the coefficients and energy values corresponding to each MO, thus formed. For this, the Secular equations can be set up (available in any quantum mechanics book) corresponding to each symmetry as follows:

$$A_u \ Symmetry: \begin{vmatrix} \alpha - E & \beta \\ \beta & \alpha - \beta - E \end{vmatrix} = 0$$

$$B_{1u} \ Symmetry: \begin{vmatrix} \alpha - E & \beta & 2\beta \\ \beta & \alpha + \beta - E & 0 \\ 2\beta & 0 & \alpha + \beta - E \end{vmatrix} = 0$$

$$B_{2g} \ Symmetry: \begin{vmatrix} \alpha - E & \beta \\ \beta & \alpha + \beta - E \end{vmatrix} = 0$$

$$B_{3g} \ Symmetry: \begin{vmatrix} \alpha - E & \beta & 2\beta \\ \beta & \alpha - \beta - E & 0 \\ 2\beta & 0 & \alpha - \beta - E \end{vmatrix} = 0$$

If we now expand each determinant into polynomial equation and taking the value of α as zero level and β as unit of energy (unity), we get the energy values of different MOs and subsequent solutions of those give the values of coefficients. The relative energies, thus found, with their orbital symmetries are given below in Figure 11.24.

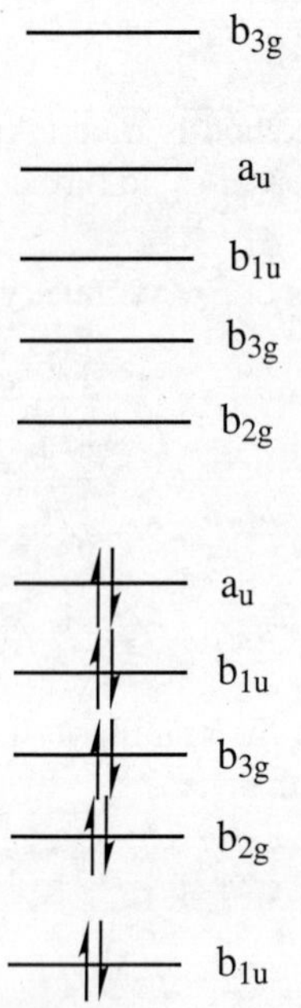

Figure 11.24: π-MO diagram of Naphthalene molecule.

From this MO diagram and its electronic configuration, it is pretty clear that lowest electronic transition corresponds to promotion of electron from a_u to b_{2g} orbital. Other lower energy excited states are formed from excitation of electrons from a_u to b_{3g} or b_{1u} to b_{2g} orbital *etc*. The ground state electronic configuration of naphthalene is $(b_{1u})^2(b_{2g})^2(b_{3g})^2(b_{1u})^2(a_u)^2$ with a totally symmetric representation of A_g. Already we have learnt how to find the symmetry of excited state from its electronic configuration. The electronic configurations corresponding to different lower energy excited states are shown below and also from D_{2h} character table their respective symmetries are determined from the direct product of the symmetries of two occupied orbitals. The probable excited states are:

$(b_{1u})^2(b_{2g}^2(b_{3g})^2(b_{1u})^2(a_u)^1(b_{2g})^1$; Symmetry: $a_u \times b_{2g} = B_{2u}$

$(b_{1u})^2(b_{2g})^2(b_{3g})^2(b_{1u})^2(a_u)^1(b_{3g})^1$; Symmetry: $a_u \times b_{3g} = B_{3u}$

$(b_{1u})^2(b_{2g})^2(b_{3g})^2(b_{1u})^1(a_u)^2(b_{2g})^1$; Symmetry: $b_{1u} \times b_{2g} = B_{3u}$

Hence, it is found that few lower energy transitions correspond to $A_g \rightarrow B_{2u}$ or two $A_g \rightarrow B_{3u}$ type. At this point, let us check which transition is allowed and which are forbidden from calculation of transition moment integrals, (**M**)= $\langle \psi_i|\mu|\psi_j \rangle$. Here, ψ_i is the wave function of initial or originating state, ψ_f is of final or destination state and dipole moment, $\mu = \mu_x + \mu_y + \mu_z$ and they correspond to B_{3u}, B_{2u} and B_{1u} respectively, for three components, obtained from D_{2h} character table.

For $A_g \to B_{2u}$ transition:

$$M_x = \langle \psi_{gr}|\mu_x|\psi_{ex}\rangle = A_g \times B_{3u} \times B_{2u} = B_{1g};\ \text{Hence}, M_x = 0$$

$$M_y = \langle \psi_{gr}|\mu_y|\psi_{ex}\rangle = A_g \times B_{2u} \times B_{2u} = A_g;\ \text{Hence}, M_y \neq 0$$

$$M_z = \langle \psi_{gr}|\mu_z|\psi_{ex}\rangle = A_g \times B_{1u} \times B_{2u} = B_{3g};\ \text{Hence}, M_z = 0$$

Hence, it is clear that this transition is allowed due to the y component of dipole moment and hence, the transition is y- or short axis polarized.

For $A_g \to B_{3u}$ transition:

$$M_x = \langle \psi_{gr}|\mu_x|\psi_{ex}\rangle = A_g \times B_{3u} \times B_{3u} = A_g;\ \text{Hence}, M_x \neq 0$$

$$M_y = \langle \psi_{gr}|\mu_y|\psi_{ex}\rangle = A_g \times B_{2u} \times B_{3u} = B_{1g};\ \text{Hence}, M_y = 0$$

$$M_z = \langle \psi_{gr}|\mu_z|\psi_{ex}\rangle = A_g \times B_{1u} \times B_{3u} = B_{2g};\ \text{Hence}, M_z = 0$$

Therefore, the two similar symmetry transitions between A_g state and B_{3u} state (corresponding to orbitals, a_u to b_{3g} and b_{1u} to b_{2g}) are allowed and x-polarized or long axis polarized.

Here, one interesting point must be noted that although, from the energy diagram of orbitals it is evident that the two $A_g \to B_{3u}$ transitions (corresponding to orbitals, a_u to b_{3g} and b_{1u} to b_{2g}) have the same energy, but experimental results show that there is a deviation between these two transitions which are presented in the Table 11.22.

The deviation of experimental result from the calculated result can be explained from *configuration interaction* (CI), which is basically the way two excited states (here two B_{3u} states) interact and get splitted into two widely spaced B_{3u} states of different energy as shown in Figure 11.25. At this point it is worthwhile to mention that instead of interaction between two electronic states, Fermi resonance is an example of CI between two vibrational states.

Table 11.22: Experimentally observed observed different electronic transitions from ground A_g state in Naphthalene.

Transition	*Energy* (cm^{-1})	*Polarization*
$A_g \to B_{3u}$	31800	x-polarization
$A_g \to B_{2u}$	34700	y-polarization
$A_g \to B_{3u}$	31800	x-polarization

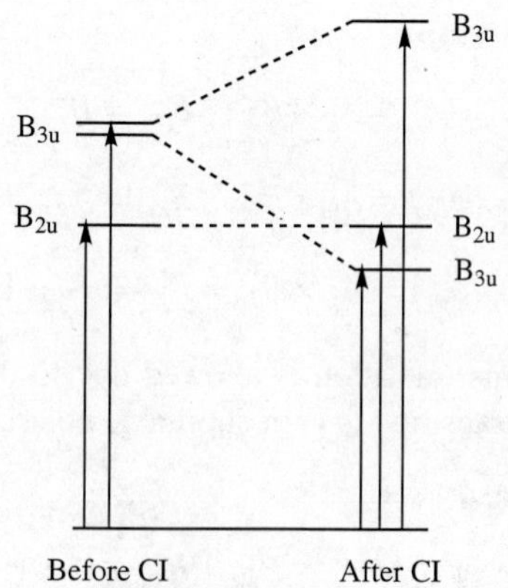

Figure 11.25: Energy level diagram of before and after configuration interaction (CI) of two B_{3u} states.

EXERCISES

11.1 Using projection operation technique, find the SALCs of water molecule and their corresponding symmetries as well. How can you check whether the constructed SALCs are genuine?

11.2 For BCl_3 molecule, using projection operation method, show that during all possible vibrations, the overall contribution of each bond are equal.

11.3 In BCl_3 molecule, show that vibrations are orthogonal to each other.

11.4 For a D_{4h} symmetry molecule, show that there is a doubly degenerate mode of vibration.

11.5 Find the symmetries of different vibrations of methane molecule. Show pictorial representations of all possible vibrations.

11.6 Find the symmetries of the SALCs of all possible bond stretching vibrations of an octahedron.

11.7 From projection operation method, find the MOs of *trans*-1,3 butadiene and write down the ground state electronic configuration as well.

11.8 Deduce the pictorial representations of the SALCs of benzene molecule and from there calculate the number of nodes present in each π-MO of benzene.

11.9 Show that the π-bond order of two adjacent C-atoms in benzene are 0.5.

11.10 Determine the symmetries of various SALCs for π-bonding in naphthalene molecule. Also find the SALC associated with A_u symmetry.

11.11 Show how, due to configuration interaction, the energy levels reorient themselves for naphthalene molecule?

11.12 Show that in naphthalene, there is no such electronic transition whose polarization is perpendicular to the plane of the molecule.

Chapter 12

Introduction to Stereographic Projection

12.1 Introduction

This is a method which is used to find the complete list of symmetry operations of a given point group from its *Schenflies notation*. Schenflies notation indicates only the basic minimum symmetry elements present in a point group. For example, D_{4h} point group symbol, simply gives information that the principal axis is C_4, there exists horizontal plane, there are four C_2 axes present which are perpendicular to the principal axis *etc.*, although, this point group possesses other symmetry elements, like, inversion centre, improper rotational axis, vertical planes.

To construct a stereographic projection for any point group, the convention is to first imagine a three-dimensional sphere onto which all the symmetry elements of the point group are placed. The main idea is that any point on the surface of the sphere is projected onto a common plane—typically the equatorial plane, which divides the sphere into the upper and lower hemispheres. The effects of symmetry operations on these points, corresponding to a given molecule, are then observed and analyzed on this plane. Few general rules or conventions are followed for construction of stereographic projection:

Rule1: Any point, situated at the upper hemisphere, (say, point at P) and projected on the equatorial plane, (+) sign is used for indicating that point and (O) sign is used to indicate the point if original position of point is below the equatorial plane or lower hemisphere (say, R).

Imagine a three dimensional sphere and the equatorial plane (shown in shaded area in Figure 12.1) divides the sphere into two hemisphere or lobes. Both the points, P and R are projected at the same position on the equatorial plane (point Q). To distinguish the origin of the point, Q, (+) and (O) notations are used for upper hemisphere and lower hemisphere respectively.

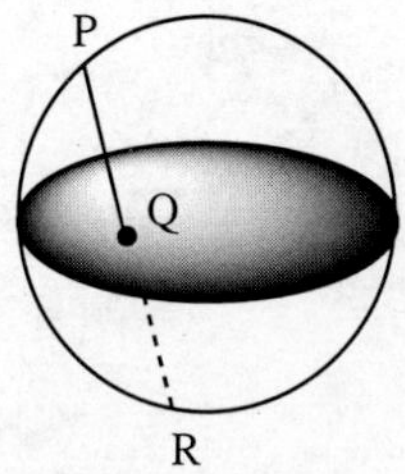

Figure 12.1: Projection on equatorial plane from surface of a sphere.

Rule 2: The equatorial plane is represented by a circle of solid line, if σ_h is present in that point group and represented by dotted line, if σ_h is absent as shown in Figure 12.2.

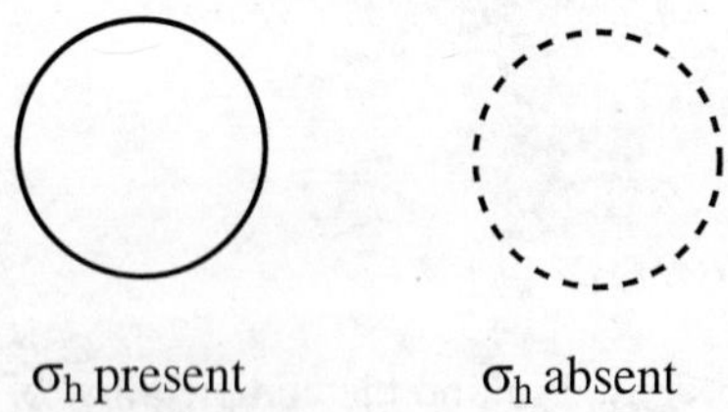

Figure 12.2: Representation of equatorial plane in presence and absence of σ_h plane.

From Schoenflies notations, it is evident whether a molecule possess any horizontal plane or not. If it starts with notation C, indicates that there is no horizontal plane and presence of horizontal plane is indicated by symbol, D.

Rule 3: The principal axis, present in a molecule, perpendicular to the equatorial plane, is placed at the centre of the equatorial plane and denoted by the following notations, presented in Figure 12.3.

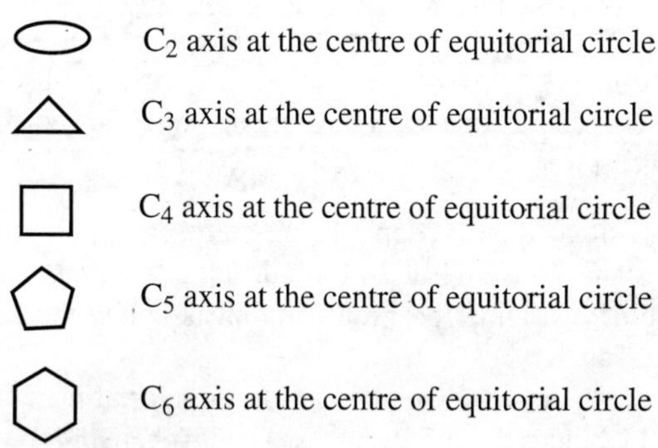

Figure 12.3: Representation of different principal axis.

Rule 4: The presence of a vertical symmetry plane is shown by a solid line along the diameter of the equatorial plane. Molecules may or may not possess horizontal plane; both are shown in Figure 12.4.

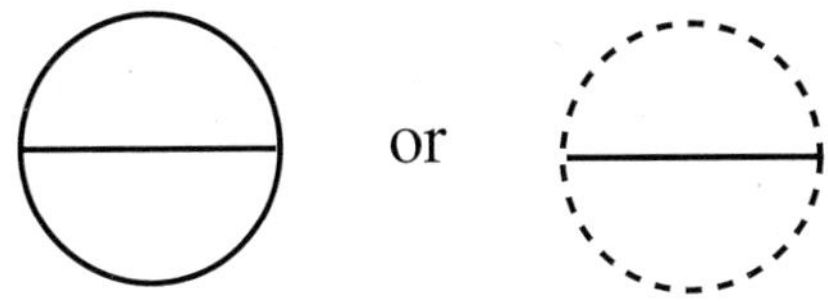

Figure 12.4: Representation of vertical plane.

Rule 5: C_2 axis, perpendicular to the principal axis, is denoted by a dotted line along diameter with two pin heads at the two ends (shown in Figure 12.5).

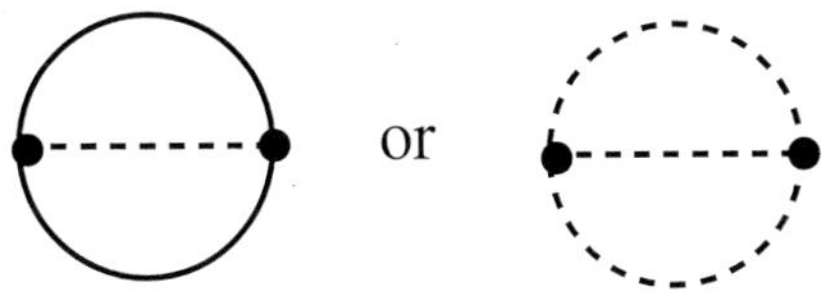

Figure 12.5: Representation of C_2 axis, perpendicular to the principal axis.

Let us try to develop stereographic projection taking few examples:

12.2 Example 1 (C_{3v} Point Group)

From the Schoenflies notation the molecule has C_3 principal axis, no horizontal plane and 3 vertical planes and no C_2 axis perpendicular to the principal axis. With these information and following the above mentioned conventions, the stereographic projection will be as shown in Figure 12.6.

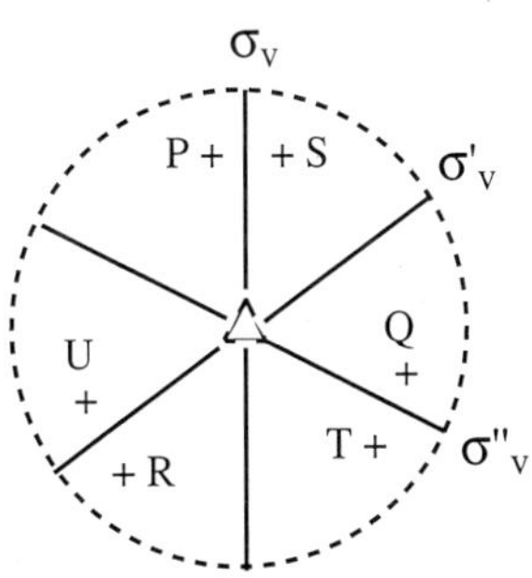

Figure 12.6: Stereographic projection of C_{3v} point group.

Consider an object P, above the equatorial plane and hence the point has been denoted by (+) sign. Now the effect of the symmetry operations, obtained from Schoenflies notation, on the point is observed. Here, basically we have to perform E, C_3, C_3^2 and three σ_v on the point, P. From the symmetry operations, C_3 and C_3^2, the two points, Q and R are obtained, respectively. Since, after these operations, points are not going to change its hemisphere or lobe, hence in both the cases they are denoted by (+) sign. After the operations of three σ_v's, as shown in Figure 12.6, the newly generated points would be at S, T and U with retention of (+) sign indicating no change over of lobe in the sphere. Hence, we can see that there are total six points in the stereographic projection, indicating the total number of symmetry operations as six. If we start from point, P and try to reach any other point, we would see that only these six symmetry operations will help to reach that point, no other symmetry operations are required for the same.

For example,

$P(+) \rightarrow P(+)$; This can be reached using operation E

$P(+) \rightarrow S(+)$; This can be reached using operation σ_v

$P(+) \rightarrow Q(+)$; This can be reached using operation C_3

$P(+) \rightarrow T(+)$; This can be reached using operation σ'_v

$P(+) \rightarrow R(+)$; This can be reached using operation C_3^2

$P(+) \rightarrow U(+)$; This can be reached using operation σ''_v

Therefore, starting from point P, every other points can be reached using the symmetry operations, $E, C_3, C_3^2, \sigma_v, \sigma'_v, \sigma''_v$. Hence, this is the full list of symmetry operations, obtained from stereographic projection, corresponding to C_{3v} point group.

The binary operations and consequence construction of multiplication table can easily be done using projection operation technique. For example, we can try to carry out the binary operation, $(\sigma_v.C_3)$ starting from point, P(+). First operation, C_3 will move the point to Q(+) and consequent second operation, σ_v will bring the point to U(+). Now, if we try to move the initial point, P(+) to the final point, U(+) in a single operation, it is found to be σ''_v.

Therefore, we find that,

$$\sigma_v.C_3 = \sigma''_v$$

In a similar way, any other binary operations can be performed and a full multiplication table can be constructed for C_{3v} point group.

12.3 Example 2 (D_{3d} Point Group)

Schenflies notation indicates the presence of a principal C_3 axis, three C_2 axes perpendicular to the principal axis and three dihedral planes bisecting the angles between those C_2 axes. Like the previous example, the symmetry operations obtained from the Schoenflies notation, would be applied to generate different points on the equatorial plane.

Let us start with a point, P_1 at the upper lobe of the sphere. In Figure 12.7, it is seen that the points have been denoted by either (+) or (O) signs, which indicate that after some symmetry operation, the point shift from upper lobe to lower lobe of the sphere.

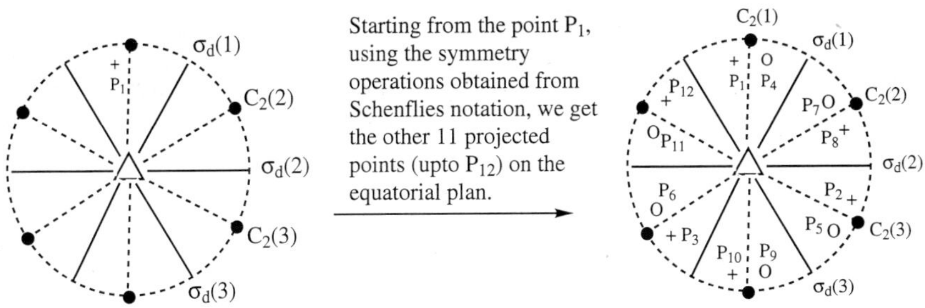

Figure 12.7: Stereographic projection of D_{3d} point group.

Since there are twelve projection points, hence there would be total twelve symmetry operations in D_{3d} point group. Out of those twelve operations, nine operations can be traced from the Schoenflies notation; $E, C_3, C_3^2, C_2(1), C_2(2), C_2(3), \sigma_d(1)$, $\sigma_d(2)$ and $\sigma_d(3)$. The other three operations can be found as follows:

Let us ask the question, how to reach the point P_9(O) from P_1(+)? Or P_7(O) from P_3(+)? Or P_{11}(O) from P_2(+), and so on. If we see carefully, it is evident that these kind of pairs of points can be inter-transformable via inversion operation (i), no other alternative symmetry operation can do the same job. Final or destination point can be reached from the initial point by travelling exactly the same distance but in opposite direction. In every case, the point changes from upper hemisphere to lower hemisphere. Therefore, it can be concluded that **inversion element (i) is present** in D_{3d} point group.

Next, we can examine, how to reach point P_7(O) from P_1(+)? Or P_9(O) from P_2(+)? And so on. Those can be achieved by rotation 60 degree through C_6 axis followed by reflection on the equatorial plane which is perpendicular to the principal axis. Actually, both the symmetry operations are non-existent in this point group. The first step would take the point to P_7(+) (This point do not basically exist in the diagram) and consequent reflection on imaginary σ_h plane would bring the point to P_7(O). That means **there is S_6 symmetry element present** in the point group. We can check the other possible symmetry operations generated by S_6 symmetry element. The full list of symmetry operations are: $S_6^1, S_6^2(=C_3), S_6^3(=i), S_6^4(=C_3^2), S_6^5, S_6^6(=E)$.

Hence, effectively, S_6 symmetry element can generate two operations, S_6 and S_6^5.

One point should be noted that since there is S_6 element present in the point group, the symbol in the center of the diagram, Δ must be replaced by $\blacktriangle$ symbol.

Therefore, ultimately, we get total twelve symmetry operations:

$$E, C_3, C_3^2, C_2(1), C_2(2), C_2(3), \sigma_d(1), \sigma_d(2), \sigma_d(3), S_6 \text{ and } S_6^5$$

Hence, what we find that starting from one initial point, we will try to reach all other points present in the diagram and would get the corresponding symmetry operations one by one.

As we have done in the previous example, although not repeating the same task here, we can try binary symmetry operation and generate full multiplication table for D_{3d}.

12.4 Example 3 (D_{3h} Point Group)

Schoenflies notation indicates the presence of C_3 as principal axis, three C_2 axes perpendicular to the principal, C_3 axis and σ_h plane. Hence, from the Schoenflies notation, the following seven symmetry operations can be identified: $E, C_3, C_3^2, C_2(1)$, $C_2(2), C_2(3)$ and σ_h. As we have done earlier, start with a point P_1 at the upper hemisphere and its projection is represented by (+) symbol. Since, σ_h plane is present here, hence, each (+) point in the diagram would be reflected by the equatorial plane and hence for each (+) point there would be a point at the same position but in lower hemisphere and would be designated by (O) symbol and vice versa. Therefore, each P will carry two points, one at upper hemisphere, with (+) symbol and another one with (O) symbol at lower hemisphere as shown in Figure 12.8.

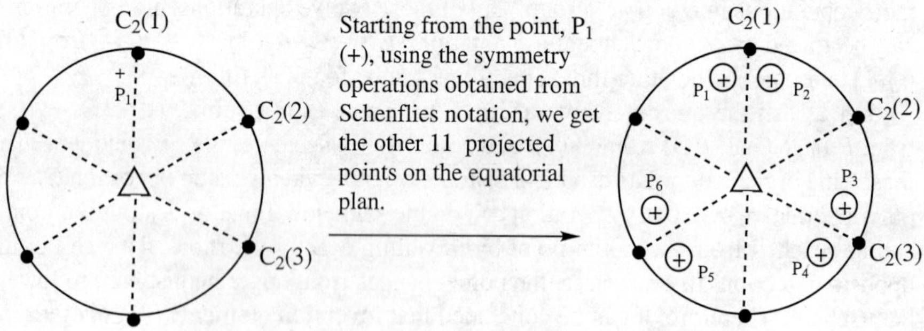

Figure 12.8: Stereographic projection of D_{3h} point group.

We have find another five symmetry operations which can not be traced from Schoenflies notation.

Let us check the transformation from initial point, P_1(+) to P_3(O) or P_2(+) to P_4(O) and so on. These points can only be reachable through S_3 symmetry element and one S_3 element can generate two operations, namely, S_3^1 and S_3^5.

Similarly, the transformation between P_1(+) and P_2(+) or P_1(O) and P_2(O) is possible if there is any σ_v plane present coincident with $C_2(1)$ axis. Similarly there would be other two σ_v planes along the other two C_2 axes. Since, there are three σ_v planes present along the three C_2 axes, the dotted lines would have to be replaced by solid lines and the final form of the diagram would be as shown in Figure 12.9 and the complete list of symmetry operations corresponding to D_{3h} poit group would be:

$$E, C_3, C_3^2, C_2(1), C_2(2), C_2(3), \sigma_h, S_3^1, S_3^5, \sigma_v(1), \sigma_v(2) \text{ and } \sigma_v(3)$$

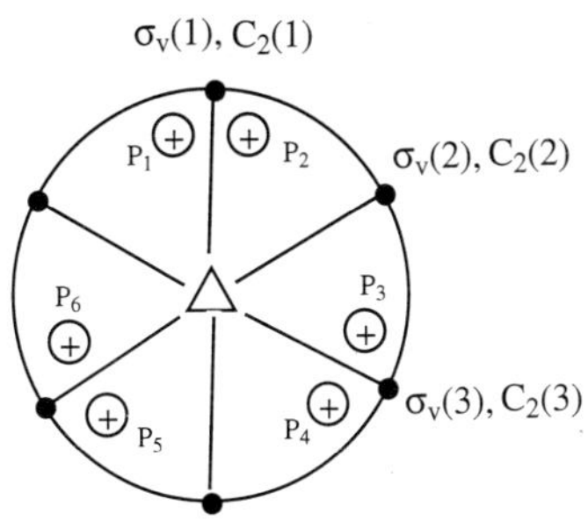

Figure 12.9: Stereographic projection of D_{3h} point group.

EXERCISES

12.1 Using stereographic projection technique, construct the multiplication table for C_{3v} point group.

12.2 Using stereographic projection technique, construct the multiplication table for D_{3d} point group.

12.3 Find the list of symmetry operations for D_{3h} point group from stereographic technique.

Appendix

Character Tables for Different Point Groups

1. The Nonaxial Groups

C_1	E
A	1

C_s	E	σ_h		
A'	1	1	x, y, R_z	x^2, y^2, z^2, xy
A''	1	−1	z, R_x, R_y	yz, xz

C_i	E	i		
A_g	1	1	R_x, R_y, R_z	x^2, y^2, z^2 xy, xz, yz
A_u	1	−1	x, y, z	

2. The C_n Groups

C_2	E	C_2		
A	1	1	z, R_z	x^2, y^2, z^2, xy
B	1	−1	x, y, R_x, R_y	yz, xz

C_3	E	C_3	C_3^2		$\varepsilon = \exp(2\pi i/3)$
A	1	1	1	z, R_z	$x^2 + y^2, z^2$
E	$\begin{cases}1 \\ 1\end{cases}$	$\begin{matrix}\varepsilon \\ \varepsilon^*\end{matrix}$	$\begin{matrix}\varepsilon^* \\ \varepsilon\end{matrix}$	$(x, y)(R_x, R_y)$	$(x^2 - y^2, xy)(yz, xz)$

C_4	E	C_4	C_2	C_4^3		
A	1	1	1	1	z, R_z	$x^2 + y^2, z^2$
B	1	−1	1	−1		$x^2 - y^2, xy$
E	$\begin{cases}1 \\ 1\end{cases}$	$\begin{matrix}i \\ -i\end{matrix}$	$\begin{matrix}-1 \\ -1\end{matrix}$	$\begin{matrix}-i \\ i\end{matrix}$	$(x, y)(R_x, R_y)$	(yz, xz)

C_5	E	C_5	C_5^2	C_5^3	C_5^4		$\varepsilon = \exp(2\pi i/5)$
A	1	1	1	1	1	z, R_z	$x^2 + y^2, z^2$
E_1	1	ε	ε^2	ε^{2*}	ε^*	$(x, y)\ (R_x, R_y)$	(yz, xz)
	1	ε^*	ε^{2*}	ε^2	ε		
E_2	1	ε^2	ε^*	ε	ε^{2*}		$(x^2 - y^2, xy)$
	1	ε^{2*}	ε	ε^*	ε^2		

C_6	E	C_6	C_3	C_2	C_3^2	C_6^5		$\varepsilon = \exp(2\pi i/6)$
A	1	1	1	1	1	1	z, R_z	$x^2 + y^2, z^2$
B	1	−1	1	−1	1	−1		
E_1	1	ε	$-\varepsilon^*$	−1	$-\varepsilon$	ε^*	(x, y)	(xz, yz)
	1	ε^*	$-\varepsilon$	−1	$-\varepsilon^*$	ε	(R_x, R_y)	
E_2	1	$-\varepsilon^*$	$-\varepsilon$	1	$-\varepsilon^*$	$-\varepsilon$		$(x^2 - y^2, xy)$
	1	$-\varepsilon$	$-\varepsilon^*$	1	$-\varepsilon$	$-\varepsilon^*$		

C_7	E	C_7	C_7^2	C_7^3	C_7^4	C_7^5	C_7^6		$\varepsilon = \exp(2\pi i/7)$
A	1	1	1	1	1	1	1	z, R_z	$x^2 + y^2, z^2$
E_1	1	ε	ε^2	ε^3	ε^{3*}	ε^{2*}	ε^*	(x, y)	(xz, yz)
	1	ε^*	ε^{2*}	ε^{3*}	ε^3	ε^2	ε	(R_x, R_y)	
E_2	1	ε^2	ε^{3*}	ε^*	ε	ε^3	ε^{2*}		$(x^2 - y^2, xy)$
	1	ε^{2*}	ε^3	ε	ε^*	ε^{3*}	ε^2		
E_3	1	ε^3	ε^*	ε^2	ε^{2*}	ε	ε^{3*}		
	1	ε^{3*}	ε	ε^{2*}	ε^2	ε^*	ε^3		

C_8	E	C_8	C_4	C_2	C_4^3	C_8^3	C_8^5	C_8^7		$\varepsilon = \exp(2\pi i/8)$
A	1	1	1	1	1	1	1	1	z, R_z	$x^2 + y^2, z^2$
B	1	−1	1	1	1	−1	−1	−1		
E_1	1	ε	i	−1	$-i$	$-\varepsilon^*$	$-\varepsilon$	ε^*	(x, y)	(xz, yz)
	1	ε^*	$-i$	−1	i	$-\varepsilon$	$-\varepsilon^*$	ε	(R_x, R_y)	
E_2	1	i	−1	1	−1	$-i$	i	$-i$		$(x^2 - y^2, xy)$
	1	$-i$	−1	1	−1	i	$-i$	i		
E_3	1	$-\varepsilon$	i	−1	$-i$	ε^*	ε	$-\varepsilon^*$		
	1	$-\varepsilon^*$	$-i$	−1	i	ε	ε^*	$-\varepsilon$		

3. The D_n Groups

D_2	E	$C_2(z)$	$C_2(y)$	$C_2(x)$		
A	1	1	1	1		x^2, y^2, z^2
B_1	1	1	−1	−1	z, R_z	xy
B_2	1	−1	1	−1	y, R_y	xz
B_3	1	−1	−1	1	x, R_x	yz

D_3	E	$2C_3$	$3C_2$		
A_1	1	1	1		x^2+y^2, z^2
A_2	1	1	–1	z, R_z	
E	2	–1	0	$(x, y)(R_x, R_y)$	$(x^2-y^2, xy)(xz, yz)$

D_4	E	$2C_4$	$C_2(=C_4^2)$	$2C_2'$	$2C_2''$		
A_1	1	1	1	1	1		x^2+y^2, z^2
A_2	1	1	1	–1	–1	z, R_z	
B_1	1	–1	1	1	–1		x^2-y^2
B_2	1	–1	1	–1	1		xy
E	2	0	–2	0	0	$(x, y)(R_x, R_y)$	(xz, yz)

D_5	E	$2C_5$	$2C_5^2$	$5C_2$		
A_1	1	1	1	1		x^2+y^2, z^2
A_2	1	1	1	–1	z, R_z	
E_1	2	2 cos 72°	2 cos 144°	0	$(x, y)(R_x, R_y)$	(xz, yz)
E_2	2	2 cos 144°	2 cos 72°	0		(x^2-y^2, xy)

D_6	E	$2C_6$	$2C_3$	C_2	$3C_2'$	$3C_2''$		
A_1	1	1	1	1	1	1		x^2+y^2, z^2
A_2	1	1	1	1	–1	–1	z, R_z	
B_1	1	–1	1	–1	1	–1		
B_2	1	–1	1	–1	–1	1		
E_1	2	1	–1	–2	0	0	$(x, y)(R_x, R_y)$	(xz, yz)
E_2	2	–1	–1	2	0	0		(x^2-y^2, xy)

4. The C_{nv} Groups

C_{2v}	E	C_2	$\sigma_v(xz)$	$\sigma_v'(yz)$		
A_1	1	1	1	1	z	x^2, y^2, z^2
A_2	1	1	–1	–1	R_z	xy
B_1	1	–1	1	–1	x, R_y	xz
B_2	1	–1	–1	1	y, R_x	yz

C_{3v}	E	$2C_3$	$3\sigma_v$		
A_1	1	1	1	z	x^2+y^2, z^2
A_2	1	1	–1	R_z	
E	2	–1	0	$(x, y)(R_x, R_y)$	$(x^2-y^2, xy)(xz, yz)$

C_{4v}	E	$2C_4$	C_2	$2\sigma_v$	$2\sigma_d$		
A_1	1	1	1	1	1	z	x^2+y^2, z^2
A_2	1	1	1	−1	−1	R_z	
B_1	1	−1	1	1	−1		x^2-y^2
B_2	1	−1	1	−1	1		xy
E	2	0	−2	0	0	(x, y) (R_x, R_y)	(xz, yz)

C_{5v}	E	$2C_5$	$2C_5^2$	$5\sigma_v$		
A_1	1	1	1	1	z	x^2+y^2, z^2
A_2	1	1	1	−1	R_z	
E_1	2	2 cos 72°	2 cos 144°	0	(x, y) (R_x, R_y)	(xz, yz)
E_2	2	2 cos 144°	2 cos 72°	0		(x^2-y^2, xy)

C_{6v}	E	$2C_6$	$2C_3$	C_2	$3\sigma_v$	$3\sigma_d$		
A_1	1	1	1	1	1	1	z	x^2+y^2, z^2
A_2	1	1	1	1	−1	−1	R_z	
B_1	1	−1	1	−1	1	−1		
B_2	1	−1	1	−1	−1	1		
E_1	2	1	−1	−2	0	0	(x, y) (R_x, R_y)	(xz, yz)
E_2	2	−1	−1	2	0	0		(x^2-y^2, xy)

5. The C_{nh} Groups

C_{2h}	E	C_2	i	σ_h		
A_g	1	1	1	1	R_z	x^2, y^2, z^2, xy
B_g	1	−1	1	−1	R_x, R_y	xz, yz
A_u	1	1	−1	−1	z	
B_u	1	−1	−1	1	x, y	

C_{3h}	E	C_3	C_3^2	σ_h	S_3	S_3^5		$\varepsilon = \exp(2\pi i/3)$
A'	1	1	1	1	1	1	R_z	x^2+y^2, z^2
E'	$\left\{\begin{matrix}1\\1\end{matrix}\right.$	$\begin{matrix}\varepsilon\\\varepsilon^*\end{matrix}$	$\begin{matrix}\varepsilon^*\\\varepsilon\end{matrix}$	$\begin{matrix}1\\1\end{matrix}$	$\begin{matrix}\varepsilon\\\varepsilon^*\end{matrix}$	$\left.\begin{matrix}\varepsilon^*\\\varepsilon\end{matrix}\right\}$	(x, y)	(x^2-y^2, xy)
A''	1	1	1	−1	−1	−1	z	
E''	$\left\{\begin{matrix}1\\1\end{matrix}\right.$	$\begin{matrix}\varepsilon\\\varepsilon^*\end{matrix}$	$\begin{matrix}\varepsilon^*\\\varepsilon\end{matrix}$	$\begin{matrix}-1\\-1\end{matrix}$	$\begin{matrix}-\varepsilon\\-\varepsilon^*\end{matrix}$	$\left.\begin{matrix}-\varepsilon^*\\-\varepsilon\end{matrix}\right\}$	(R_x, R_y)	(xz, yz)

C_{4h}	E	C_4	C_2	C_4^3	i	S_4^3	σ_h	S_4		
A_g	1	1	1	1	1	1	1	1	R_z	x^2+y^2, z^2
B_g	1	–1	1	–1	1	–1	1	–1		x^2-y^2, xy
E_g	1	i	–1	$-i$	1	i	–1	$-i$	(R_x, R_y)	(xz, yz)
	1	$-i$	–1	i	1	$-i$	–1	i		
A_u	1	1	1	1	–1	–1	–1	–1	z	
B_u	1	–1	1	–1	–1	1	–1	1		
E_u	1	i	–1	$-i$	–1	$-i$	1	i	(x, y)	
	1	$-i$	–1	i	–1	i	1	$-i$		

C_{5h}	E	C_5	C_5^2	C_5^3	C_5^4	σ_h	S_5	S_5^7	S_5^3	S_5^9		$\varepsilon = \exp(2\pi i/5)$
A'	1	1	1	1	1	1	1	1	1	1	R_z	x^2+y^2, z^2
E_1'	1	ε	ε^2	ε^{2*}	ε^*	1	ε	ε^2	ε^{2*}	ε^*	(x, y)	
	1	ε^*	ε^{2*}	ε^2	ε	1	ε^*	ε^{2*}	ε^2	ε		
E_2'	1	ε^2	ε^*	ε	ε^{2*}	1	ε^2	ε^*	ε	ε^{2*}		(x^2-y^2, xy)
	1	ε^{2*}	ε	ε^*	ε^2	1	ε^{2*}	ε	ε^*	ε^2		
A''	1	1	1	1	1	–1	–1	–1	–1	–1	z	
E_1''	1	ε	ε^2	ε^{2*}	ε^*	–1	$-\varepsilon$	$-\varepsilon^2$	$-\varepsilon^{2*}$	$-\varepsilon^*$	(R_x, R_y)	(xz, yz)
	1	ε^*	ε^{2*}	ε^2	ε	–1	$-\varepsilon^*$	$-\varepsilon^{2*}$	$-\varepsilon^2$	$-\varepsilon$		
E_2''	1	ε^2	ε^*	ε	ε^{2*}	–1	$-\varepsilon^2$	$-\varepsilon^*$	$-\varepsilon$	$-\varepsilon^{2*}$		
	1	ε^{2*}	ε	ε^*	ε^2	–1	$-\varepsilon^{2*}$	$-\varepsilon$	$-\varepsilon^*$	$-\varepsilon^2$		

C_{6h}	E	C_6	C_3	C_2	C_3^2	C_6^5	i	S_3^5	S_6^5	σ_h	S_6	S_3		$\varepsilon = \exp(2\pi i/6)$
A_g	1	1	1	1	1	1	1	1	1	1	1	1	R_z	x^2+y^2, z^2
B_g	1	–1	1	–1	1	–1	1	–1	1	–1	1	–1		
E_{1g}	1	ε	$-\varepsilon^*$	–1	$-\varepsilon$	ε^*	1	ε	$-\varepsilon^*$	–1	$-\varepsilon$	ε^*	(R_x, R_y)	(xz, yz)
	1	ε^*	$-\varepsilon$	–1	$-\varepsilon^*$	ε	1	ε^*	$-\varepsilon$	–1	$-\varepsilon^*$	ε		
E_{2g}	1	$-\varepsilon^*$	$-\varepsilon$	1	$-\varepsilon^*$	$-\varepsilon$	1	$-\varepsilon^*$	$-\varepsilon$	1	$-\varepsilon^*$	$-\varepsilon$		(x^2-y^2, xy)
	1	$-\varepsilon$	$-\varepsilon^*$	1	$-\varepsilon$	$-\varepsilon^*$	1	$-\varepsilon$	$-\varepsilon^*$	1	$-\varepsilon$	$-\varepsilon^*$		
A_u	1	1	1	1	1	1	–1	–1	–1	–1	–1	–1	z	
B_u	1	–1	1	–1	1	–1	–1	1	–1	1	–1	1		
E_{1u}	1	ε	$-\varepsilon^*$	–1	$-\varepsilon$	ε^*	–1	$-\varepsilon$	ε^*	1	ε	$-\varepsilon^*$	(x, y)	
	1	ε^*	$-\varepsilon$	–1	$-\varepsilon^*$	ε	–1	$-\varepsilon^*$	ε	1	ε^*	$-\varepsilon$		
E_{2u}	1	$-\varepsilon^*$	$-\varepsilon$	1	$-\varepsilon^*$	$-\varepsilon$	–1	ε^*	ε	–1	ε^*	ε		
	1	$-\varepsilon$	$-\varepsilon^*$	1	$-\varepsilon$	$-\varepsilon^*$	–1	ε	ε^*	–1	ε	ε^*		

6. The D_{nh} Groups

D_{2h}	E	$C_2(z)$	$C_2(y)$	$C_2(x)$	i	$\sigma(xy)$	$\sigma(xz)$	$\sigma(yz)$		
A_g	1	1	1	1	1	1	1	1		x^2, y^2, z^2
B_{1g}	1	1	−1	−1	1	1	−1	−1	R_z	xy
B_{2g}	1	−1	1	−1	1	−1	1	−1	R_y	xz
B_{3g}	1	−1	−1	1	1	−1	−1	1	R_x	yz
A_u	1	1	1	1	−1	−1	−1	−1		
B_{1u}	1	1	−1	−1	−1	−1	1	1	z	
B_{2u}	1	−1	1	−1	−1	1	−1	1	y	
B_{3u}	1	−1	−1	1	−1	1	1	−1	x	

D_{3h}	E	$2C_3$	$3C_2$	σ_h	$2S_3$	$3\sigma_v$		
A_1'	1	1	1	1	1	1		$x^2 + y^2, z^2$
A_2'	1	1	−1	1	1	−1	R_z	
E'	2	−1	0	2	−1	0	(x, y)	$(x^2 - y^2, xy)$
A_1''	1	1	1	−1	−1	−1		
A_2''	1	1	−1	−1	−1	1	z	
E''	2	−1	0	−2	1	0	(R_x, R_y)	(xz, yz)

D_{4h}	E	$2C_4$	C_2	$2C_2'$	$2C_2''$	i	$2S_4$	σ_h	$2\sigma_v$	$2\sigma_d$		
A_{1g}	1	1	1	1	1	1	1	1	1	1		$x^2 + y^2, z^2$
A_{2g}	1	1	1	−1	−1	1	1	1	−1	−1	R_z	
B_{1g}	1	−1	1	1	−1	1	−1	1	1	−1		$(x^2 - y^2)$
B_{2g}	1	−1	1	−1	1	1	−1	1	−1	1		xy
E_g	2	0	−2	0	0	2	0	−2	0	0	(R_x, R_y)	(xz, yz)
A_{1u}	1	1	1	1	1	−1	−1	−1	−1	−1		
A_{2u}	1	1	1	−1	−1	−1	−1	−1	1	1	z	
B_{1u}	1	−1	1	1	−1	−1	1	−1	−1	1		
B_{2u}	1	−1	1	−1	1	−1	1	−1	1	−1		
E_u	2	0	−2	0	0	−2	0	2	0	0	(x, y)	

D_{5h}	E	$2C_5$	$2C_5^2$	$5C_2$	σ_h	$2S_5$	$2S_5^3$	$5\sigma_v$		
A_1'	1	1	1	1	1	1	1	1		x^2+y^2, z
A_2'	1	1	1	−1	1	1	1	−1	R_z	
E_1'	2	2 cos 72°	2 cos 144°	0	2	2 cos 72°	2 cos 144°	0	(x, y)	
E_2'	2	2 cos 144°	2 cos 72°	0	2	2 cos 144°	2 cos 72°	0		(x^2-y^2, xy)
A_1''	1	1	1	1	−1	−1	−1	−1		
A_2''	1	1	1	−1	−1	−1	−1	1	z	
E_1''	2	2 cos 72°	2 cos 144°	0	−2	−2 cos 72°	−2 cos 144°	0	(R_x, R_y)	(xz, yz)
E_2''	2	2 cos 144°	2 cos 72°	0	−2	−2 cos 144°	−2 cos 72°	0		

D_{6h}	E	$2C_6$	$2C_3$	C_2	$3C_2'$	$3C_2''$	i	$2S_3$	$2S_6$	σ_h	$3\sigma_d$	$3\sigma_v$		
A_{1g}	1	1	1	1	1	1	1	1	1	1	1	1		x^2+y^2, z^2
A_{2g}	1	1	1	1	−1	−1	1	1	1	1	−1	−1	R_z	
B_{1g}	1	−1	1	−1	1	−1	1	−1	1	−1	1	−1		
B_{2g}	1	−1	1	−1	−1	1	1	−1	1	−1	−1	1		
E_{1g}	2	1	−1	−2	0	0	2	1	−1	−2	0	0	(R_x, R_y)	(xz, yz)
E_{2g}	2	−1	−1	2	0	0	2	−1	−1	2	0	0		(x^2-y^2, xy)
A_{1u}	1	1	1	1	1	1	−1	−1	−1	−1	−1	−1		
A_{2u}	1	1	1	1	−1	−1	−1	−1	−1	−1	1	1	z	
B_{1u}	1	−1	1	−1	1	−1	−1	1	−1	1	−1	1		
B_{2u}	1	−1	1	−1	−1	1	−1	1	−1	1	1	−1		
E_{1u}	2	1	−1	−2	0	0	−2	−1	1	2	0	0	(x, y)	
E_{2u}	2	−1	−1	2	0	0	−2	1	1	−2	0	0		

D_{8h}	E	$2C_8$	$2C_8^3$	$2C_4$	C_2	$4C_2'$	$4C_2''$	i	$2S_8$	$2S_8^3$	$2S_4$	σ_h	$4\sigma_d$	$4\sigma_v$		
A_{1g}	1	1	1	1	1	1	1	1	1	1	1	1	1	1		x^2+y^2, z^2
A_{2g}	1	1	1	1	1	−1	−1	1	1	1	1	1	−1	−1	R_z	
B_{1g}	1	−1	−1	1	1	1	−1	1	−1	−1	1	1	1	−1		
B_{2g}	1	−1	−1	1	1	−1	1	1	−1	−1	1	1	−1	1		
E_{1g}	2	$\sqrt{2}$	$-\sqrt{2}$	0	−2	0	0	2	$\sqrt{2}$	$-\sqrt{2}$	0	−2	0	0	(R_x, R_y)	(xz, yz)
E_{2g}	2	0	0	−2	2	0	0	2	0	0	−2	2	0	0		(x^2-y^2, xy)
E_{3g}	2	$-\sqrt{2}$	$\sqrt{2}$	0	−2	0	0	2	$-\sqrt{2}$	$\sqrt{2}$	0	−2	0	0		
A_{1u}	1	1	1	1	1	1	1	−1	−1	−1	−1	−1	−1	−1		
A_{2u}	1	1	1	1	1	−1	−1	−1	−1	−1	−1	−1	1	1	z	
B_{1u}	1	−1	−1	1	1	1	−1	−1	1	1	−1	−1	−1	1		
B_{2u}	1	−1	−1	1	1	−1	1	−1	1	1	−1	−1	1	−1		
E_{1u}	2	$\sqrt{2}$	$-\sqrt{2}$	0	−2	0	0	−2	$-\sqrt{2}$	$\sqrt{2}$	0	2	0	0	(x, y)	
E_{2u}	2	0	0	−2	2	0	0	−2	0	0	2	−2	0	0		
E_{3u}	2	$-\sqrt{2}$	$\sqrt{2}$	0	−2	0	0	−2	$\sqrt{2}$	$-\sqrt{2}$	0	2	0	0		

7. The D_{nd} Groups

D_{2d}	E	$2S_4$	C_2	$2C_2'$	$2\sigma_d$		
A_1	1	1	1	1	1		x^2+y^2, z^2
A_2	1	1	1	−1	−1	R_z	
B_1	1	−1	1	1	−1		x^2-y^2
B_2	1	−1	1	−1	1	z	xy
E	2	0	−2	0	0	(x, y); (R_x, R_y)	(xz, yz)

D_{3d}	E	$2C_3$	$3C_2$	i	$2S_6$	$3\sigma_d$		
A_{1g}	1	1	1	1	1	1		x^2+y^2, z^2
A_{2g}	1	1	−1	1	1	−1	R_z	
E_g	2	−1	0	2	−1	0	(R_x, R_y)	$(x^2-y^2, xy), (xz, yz)$
A_{1u}	1	1	1	−1	−1	−1		
A_{2u}	1	1	−1	−1	−1	1	z	
E_u	2	−1	0	−2	1	0	(x, y)	

D_{4d}	E	$2S_8$	$2C_4$	$2S_8^3$	C_2	$4C_2'$	$4\sigma_d$		
A_1	1	1	1	1	1	1	1		x^2+y^2, z^2
A_2	1	1	1	1	1	−1	−1	R_z	
B_1	1	−1	1	−1	1	1	−1		
B_2	1	−1	1	−1	1	−1	1	z	
E_1	2	$\sqrt{2}$	0	$-\sqrt{2}$	−2	0	0	(x, y)	
E_2	2	0	−2	0	2	0	0		(x^2-y^2, xy)
E_3	2	$-\sqrt{2}$	0	$\sqrt{2}$	−2	0	0	(R_x, R_y)	(xz, yz)

D_{5d}	E	$2C_5$	$2C_5^2$	$5C_2$	i	$2S_{10}^3$	$2S_{10}$	$5\sigma_d$		
A_{1g}	1	1	1	1	1	1	1	1		x^2+y^2, z^2
A_{2g}	1	1	1	−1	1	1	1	−1	R_z	
E_{1g}	2	2 cos 72°	2cos 144°	0	2	2 cos 72°	2 cos 144°	0	(R_x, R_y)	(xz, yz)
E_{2g}	2	2 cos 144°	2 cos 72°	0	2	2 cos 144°	2 cos 72°	0		(x^2-y^2, xy)
A_{1u}	1	1	1	1	−1	−1	−1	−1		
A_{2u}	1	1	1	−1	−1	−1	−1	1	z	
E_{1u}	2	2 cos 72°	2 cos 144°	0	−2	−2 cos 72°	−2 cos 144°	0	(x, y)	
E_{2u}	2	2 cos 144°	2 cos 72°	0	−2	−2 cos 144°	−2 cos 72°	0		

D_{6d}	E	$2S_{12}$	$2C_6$	$2S_4$	$2C_3$	$2S_{12}^5$	C_2	$6C_2'$	$6\sigma_d$		
A_1	1	1	1	1	1	1	1	1	1		x^2+y^2, z^2
A_2	1	1	1	1	1	1	1	−1	−1	R_z	
B_1	1	−1	1	−1	1	−1	1	1	−1		
B_2	1	−1	1	−1	1	−1	1	−1	1	z	
E_1	2	$\sqrt{3}$	1	0	−1	$-\sqrt{3}$	−2	0	0	(x, y)	
E_2	2	1	−1	−2	−1	1	2	0	0		(x^2-y^2, xy)
E_3	2	0	−2	0	2	0	−2	0	0		
E_4	2	−1	−1	2	−1	−1	2	0	0		
E_5	2	$-\sqrt{3}$	1	0	−1	$\sqrt{3}$	−2	0	0	(R_x, R_y)	(xy, yz)

8. The S_n Groups

S_4	E	S_4	C_2	S_4^3		
A	1	1	1	1	R_z	x^2+y^2, z^2
B	1	−1	1	−1	z	(x^2-y^2, xy)
E	$\begin{cases}1\\1\end{cases}$	$\begin{matrix}i\\-i\end{matrix}$	$\begin{matrix}-1\\-1\end{matrix}$	$\begin{matrix}-i\\i\end{matrix}$	$(x, y); (R_x, R_y)$	(xz, yz)

S_6	E	C_3	C_3^2	i	S_6^5	S_6		$\varepsilon = \exp(2\pi i/3)$
A_g	1	1	1	1	1	1	R_z	x^2+y^2, z^2
E_g	1	ε	ε^*	1	ε	ε^*	(R_x, R_y)	(x^2-y^2, xy)
	1	ε^*	ε	1	ε^*	ε		(xz, yz)
A_u	1	1	1	−1	−1	−1	z	
E_u	1	ε	ε^*	−1	$-\varepsilon$	$-\varepsilon^*$	(x, y)	
	1	ε^*	ε	−1	$-\varepsilon^*$	$-\varepsilon$		

S_8	E	S_8	C_4	S_8^3	C_2	S_8^5	C_4^3	S_8^7		$\varepsilon = \exp(2\pi i/8)$
A	1	1	1	1	1	1	1	1	R_z	x^2+y^2, z^2
B	1	−1	1	−1	1	−1	1	−1	z	
E_1	1	ε	i	$-\varepsilon^*$	−1	$-\varepsilon$	$-i$	ε^*	(x, y);	
	1	ε^*	$-i$	$-\varepsilon$	−1	$-\varepsilon^*$	i	ε	(R_x, R_y)	
E_2	1	i	−1	$-i$	1	i	−1	$-i$		(x^2-y^2, xy)
	1	$-i$	−1	i	1	$-i$	−1	i		
E_3	1	$-\varepsilon^*$	$-i$	ε	−1	ε^*	i	$-\varepsilon$		(xz, yz)
	1	$-\varepsilon$	i	ε^*	−1	ε	$-i$	$-\varepsilon^*$		

9. The Cubic Groups

T	E	$4C_3$	$4C_3^2$	$3C_2$		$\varepsilon = \exp(2\pi i/3)$
A	1	1	1	1		$x^2+y^2+z^2$
E	1	ε	ε^*	1		$(2z^2-x^2-y^2,$
	1	ε^*	ε	1		$x^2-y^2)$
T	3	0	0	−1	$(R_x, R_y, R_z); (x, y, z)$	(xy, xz, yz)

T_h	E	$4C_3$	$4C_3^2$	$3C_2$	i	$4S_6$	$4S_6^5$	$3\sigma_h$		$\varepsilon = \exp(2\pi i/3)$
A_g	1	1	1	1	1	1	1	1		$x^2+y^2+z^2$
A_u	1	1	1	1	−1	−1	−1	−1		
E_g	$\begin{cases}1\\1\end{cases}$	$\begin{matrix}\varepsilon\\\varepsilon^*\end{matrix}$	$\begin{matrix}\varepsilon^*\\\varepsilon\end{matrix}$	$\begin{matrix}1\\1\end{matrix}$	$\begin{matrix}1\\1\end{matrix}$	$\begin{matrix}\varepsilon\\\varepsilon^*\end{matrix}$	$\begin{matrix}\varepsilon^*\\\varepsilon\end{matrix}$	$\begin{rcases}1\\1\end{rcases}$		$(2z^2-x^2-y^2, x^2-y^2)$
E_u	$\begin{cases}1\\1\end{cases}$	$\begin{matrix}\varepsilon\\\varepsilon^*\end{matrix}$	$\begin{matrix}\varepsilon^*\\\varepsilon\end{matrix}$	$\begin{matrix}1\\1\end{matrix}$	$\begin{matrix}-1\\-1\end{matrix}$	$\begin{matrix}-\varepsilon\\-\varepsilon^*\end{matrix}$	$\begin{matrix}-\varepsilon^*\\-\varepsilon\end{matrix}$	$\begin{rcases}-1\\-1\end{rcases}$		
T_g	3	0	0	−1	−3	0	0	−1	(R_x, R_y, R_z)	(xz, yz, xy)
T_u	3	0	0	−1	−3	0	0	1	(x, y, z)	

T_d	E	$8C_3$	$3C_2$	$6S_4$	$6\sigma_d$		
A_1	1	1	1	1	1		$x^2+y^2+z^2$
A_2	1	1	1	−1	−1		
E	2	−1	2	0	0		$(2z^2-x^2-y^2, x^2-y^2)$
T_1	3	0	−1	1	−1	(R_x, R_y, R_z)	
T_2	3	0	−1	−1	1	(x, y, z)	(xy, xz, yz)

O	E	$6C_4$	$3C_2(=C_4^2)$	$8C_3$	$6C_2$		
A_1	1	1	1	1	1		$x^2+y^2+z^2$
A_2	1	−1	1	1	−1		
E	2	0	2	−1	0		$(2z^2-x^2-y^2, x^2-y^2)$
T_1	3	1	−1	0	−1	$(R_x, R_y, R_z); (x, y, z)$	
T_2	3	−1	−1	0	1		(xy, xz, yz)

O_h	E	$8C_3$	$6C_2$	$6C_4$	$3C_2(=C_4^2)$	i	$6S_4$	$8S_6$	$3\sigma_h$	$6\sigma_d$		
A_{1g}	1	1	1	1	1	1	1	1	1	1		$x^2+y^2+z^2$
A_{2g}	1	1	−1	−1	1	1	−1	1	1	−1		
E_g	2	−1	0	0	2	2	0	−1	2	0		$(2z^2-x^2-y^2, x^2-y^2)$
T_{1g}	3	0	−1	1	−1	3	1	0	−1	−1	(R_x, R_y, R_z)	
T_{2g}	3	0	1	−1	−1	3	−1	0	−1	1		(xz, yz, xy)
A_{1u}	1	1	1	1	1	−1	−1	−1	−1	−1		
A_{2u}	1	1	−1	−1	1	−1	1	−1	−1	1		
E_u	2	−1	0	0	2	−2	0	1	−2	0		
T_{1u}	3	0	−1	1	−1	−3	−1	0	1	1	(x, y, z)	
T_{2u}	3	0	1	−1	−1	−3	1	0	1	−1		

10. The Groups $C_{\infty v}$ and $D_{\infty h}$ for Linear Molecules

$C_{\infty v}$	E	$2C_{\infty}^{\phi}$	$\cdots$	$\infty\alpha_v$		
$A_1 \equiv \Sigma^+$	1	1	$\cdots$	1	z	$x^2 + y^2, z^2$
$A_2 \equiv \Sigma^-$	1	1	$\cdots$	-1	R_z	
$E_1 \equiv \Pi$	2	$2\cos\phi$	$\cdots$	0	$(x, y); (R_x, R_y)$	(xz, yz)
$E_2 \equiv \Delta$	2	$2\cos 2\phi$	$\cdots$	0		$(x^2 - y^2, xy)$
$E_3 \equiv \phi$	2	$2\cos 3\phi$	$\cdots$	0		
$\cdots$	$\cdots$	$\cdots$	$\cdots$	$\cdots$		

$D_{\infty h}$	E	$2C_{\infty}^{\phi}$	$\cdots$	$\infty\sigma_v$	i	$2S_{\infty}^{\phi}$	$\cdots$	∞C_2		
Σ_g^+	1	1	$\cdots$	1	1	1	$\cdots$	1		$x^2 + y^2, z^2$
Σ_g^-	1	1	$\cdots$	-1	1	1	$\cdots$	-1	R_z	
Π_g	2	$2\cos\phi$	$\cdots$	0	2	$-2\cos\phi$	$\cdots$	0	(R_x, R_y)	(xz, yz)
Δ_g	2	$2\cos 2\phi$	$\cdots$	0	2	$2\cos 2\phi$	$\cdots$	0		$(x^2 - y^2, xy)$
$\cdots$	$\cdots$	$\cdots$	$\cdots$	$\cdots$	$\cdots$	$\cdots$	$\cdots$	$\cdots$		
Σ_u^+	1	1	$\cdots$	1	-1	-1	$\cdots$	-1	z	
Σ_u^-	1	1	$\cdots$	-1	-1	-1	$\cdots$	1		
Π_u	2	$2\cos\phi$	$\cdots$	0	-2	$2\cos\phi$	$\cdots$	0	(x, y)	
Δ_u	2	$2\cos 2\phi$	$\cdots$	0	-2	$-2\cos 2\phi$	$\cdots$	0		
$\cdots$	$\cdots$	$\cdots$	$\cdots$	$\cdots$	$\cdots$	$\cdots$	$\cdots$	$\cdots$		

11. The Icosahedral Groups*

I_h	E	$12C_5$	$12C_5^2$	$20C_3$	$15C_2$	i	$12S_{10}$	$12S_{10}^3$	$20S_6$	15σ		
A_g	1	1	1	1	1	1	1	1	1	1		$x^2+y^2+z^2$
T_{1g}	3	$\frac{1}{2}(1+\sqrt{5})$	$\frac{1}{2}(1-\sqrt{5})$	0	−1	3	$\frac{1}{2}(1-\sqrt{5})$	$\frac{1}{2}(1+\sqrt{5})$	0	−1	(R_x, R_y, R_z)	
T_{2g}	3	$\frac{1}{2}(1-\sqrt{5})$	$\frac{1}{2}(1+\sqrt{5})$	0	−1	3	$\frac{1}{2}(1+\sqrt{5})$	$\frac{1}{2}(1-\sqrt{5})$	0	−1		
G_g	4	−1	−1	1	0	4	−1	−1	1	0		
H_g	5	0	0	−1	1	5	0	0	−1	1		$(2z^2-x^2-y^2, x^2-y^2, xy, yz, zx)$
A_u	1	1	1	1	1	−1	−1	−1	−1	−1		
T_{1u}	3	$\frac{1}{2}(1+\sqrt{5})$	$\frac{1}{2}(1-\sqrt{5})$	0	−1	−3	$-\frac{1}{2}(1-\sqrt{5})$	$-\frac{1}{2}(1+\sqrt{5})$	0	1	(x, y, z)	
T_{2u}	3	$\frac{1}{2}(1-\sqrt{5})$	$\frac{1}{2}(1+\sqrt{5})$	0	−1	−3	$-\frac{1}{2}(1+\sqrt{5})$	$-\frac{1}{2}(1-\sqrt{5})$	0	1		
G_u	4	−1	−1	1	0	−4	1	1	−1	0		
H_u	5	0	0	−1	1	−5	0	0	1	−1		

*For the pure rotation group I, the outlined section in the upper left is the character table; the g subscripts should be dropped and (x, y, z) assigned to the T_1 representation.

Index